VOLUME ONE HUNDRED AND EIGHTY SIX

ADVANCES IN
IMAGING AND
ELECTRON PHYSICS

EDITOR-IN-CHIEF

Peter W. Hawkes
CEMES-CNRS
Toulouse, France

VOLUME ONE HUNDRED AND EIGHTY SIX

ADVANCES IN
IMAGING AND
ELECTRON PHYSICS

Edited by

PETER W. HAWKES
CEMES-CNRS, Toulouse, France

Amsterdam • Boston • Heidelberg • London
New York • Oxford • Paris • San Diego
San Francisco • Singapore • Sydney • Tokyo

Academic Press is an imprint of Elsevier

Cover photo credit:
Niels de Jonge et al., Practical Aspects of Transmission Electron Microscopy in Liquid
Advances in Imaging and Electron Physics (2014) 186, pp. 1-38

Academic Press is an imprint of Elsevier
32 Jamestown Road, London NW1 7BY, UK
525 B Street, Suite 1800, San Diego, CA 92101-4495, USA
225 Wyman Street, Waltham, MA 02451, USA
The Boulevard, Langford Lane, Kidlington, Oxford OX5 1GB, UK

First edition 2014
Copyright © 2014 Elsevier Inc. All rights reserved.

Notices
Knowledge and best practice in this field are constantly changing. As new research and experience
broaden our understanding, changes in research methods, professional practices, or medical treatment
may become necessary.

Practitioners and researchers must always rely on their own experience and knowledge in evaluating and
using any information, methods, compounds, or experiments described herein. In using such information
or methods they should be mindful of their own safety and the safety of others, including parties for
whom they have a professional responsibility.

To the fullest extent of the law, neither the Publisher nor the authors, contributors, or editors, assume
any liability for any injury and/or damage to persons or property as a matter of products liability,
negligence or otherwise, or from any use or operation of any methods, products, instructions, or ideas
contained in the material herein.

British Library Cataloguing in Publication Data
A catalogue record for this book is available from the British Library

Library of Congress Cataloging-in-Publication Data
A catalog record for this book is available from the Library of Congress

ISBN: 978-0-12-800264-3
ISSN: 1076-5670

For information on all Academic Press publications
visit our website at http://store.elsevier.com/

Printed in the United States of America

Working together
to grow libraries in
developing countries

ELSEVIER   Book Aid International

www.elsevier.com • www.bookaid.org

CONTENTS

PREFACE

Another volume with chapters of wide interest. First, we have an account by N. de Jonge, M. Pfaff, and D.B. Peckys of an unusual form of electron microscopy, transmission electron microscopy in liquid. Microscopists, especially life scientists, have always wanted to observe specimens at the resolution offered by the electron microscope in a less hostile environment than a high vacuum. This was indeed one of the principal reasons for the construction of the first high-voltage electron microscope in Toulouse, with which Gaston Dupouy hoped to see living material in a cell containing air and water vapor. Today, we have environmental scanning electron microscopy (recently extended to perform ESEM in aberration-corrected conditions), a renewal of wet cells and liquid flow. The authors of this chapter have made major contributions to progress in this area and have distilled their expertise into this full account.

In the second chapter, J.-J. Ding and S.-C. Pei give a very detailed presentation of the linear canonical transform. This is a more general form of the familiar Fourier transform, a relative of the Wigner distribution function, which is nowadays relatively well known. It has found applications in many areas of image processing, phase retrieval, and filter design. The authors set out the basic theory and then describe several applications.

The volume concludes with a chapter on manipulation of tiny objects in the electron microscope by A.I. Denisyuk, A.V. Krasavin, F.E. Komissarenko, and I.S. Mukhin. Very small objects can be manipulated by mechanical forces or by electrostatic or magnetic forces. The authors explain how these operations are performed. The final sections, in which the use of electrostatic and magnetic fields is examined, are particularly interesting, as these methods are less familiar.

As always, I thank the authors for the trouble they have taken to make their subjects easy to follow by readers from other subject areas.

Peter Hawkes

FUTURE CONTRIBUTIONS

H.-W. Ackermann
Electron micrograph quality

J. Andersson and J.-O. Strömberg
Radon transforms and their weighted variants

S. Ando
Gradient operators and edge and corner detection

J. Angulo
Mathematical morphology for complex and quaternion-valued images

D. Batchelor
Soft x-ray microscopy

E. Bayro Corrochano
Quaternion wavelet transforms

C. Beeli
Structure and microscopy of quasicrystals

M. Berz Ed.
Femtosecond electron imaging and spectroscopy

C. Bobisch and R. Möller
Ballistic electron microscopy

F. Bociort
Saddle-point methods in lens design

K. Bredies
Diffusion tensor imaging

A. Broers
A retrospective

R.E. Burge
A scientific autobiography

A. Carroll
Reflective electron beam lithography

N. Chandra and R. Ghosh
Quantum entanglement in electron optics

A. Cornejo Rodriguez and F. Granados Agustin
Ronchigram quantification

A. Elgammal (Vol. 187)
Homeomorphic manifold analysis (HMA): untangling complex manifolds

J. Elorza
Fuzzy operators

A.R. Faruqi, G. McMullan and R. Henderson
Direct detectors

M. Ferroni
Transmission microscopy in the scanning electron microscope

R.G. Forbes
Liquid metal ion sources

A. Gölzhäuser
Recent advances in electron holography with point sources

J. Grotemeyer and T. Muskat
Time-of-flight mass spectrometry

M. Haschke
Micro-XRF excitation in the scanning electron microscope

R. Herring and B. McMorran
Electron vortex beams

M.S. Isaacson
Early STEM development

K. Ishizuka
Contrast transfer and crystal images

K. Jensen, D. Shiffler and J. Luginsland
Physics of field emission cold cathodes

U. Kaiser
The sub-Ångström low-voltage electron microcope project (SALVE)

C.T. Koch
In-line electron holography

T. Kohashi (Vol. 187)
Spin-polarized scanning electron microscopy

O.L. Krivanek
Aberration-corrected STEM

M. Kroupa
The Timepix detector and its applications

B. Lencová
Modern developments in electron optical calculations

H. Lichte
Developments in electron holography

M. Matsuya
Calculation of aberration coefficients using Lie algebra

J.A. Monsoriu
Fractal zone plates

L. Muray
Miniature electron optics and applications

M.A. O'Keefe
Electron image simulation

V. Ortalan
Ultrafast electron microscopy

D. Paganin, T. Gureyev and K. Pavlov
Intensity-linear methods in inverse imaging

M. Pap (Vol. 187)
Hyperbolic wavelets

N. Papamarkos and A. Kesidis
The inverse Hough transform

Q. Ramasse and R. Brydson
The SuperSTEM laboratory

B. Rieger and A.J. Koster
Image formation in cryo-electron microscopy

P. Rocca and M. Donelli
Imaging of dielectric objects

J. Rodenburg
Lensless imaging

J. Rouse, H.-n. Liu and E. Munro
The role of differential algebra in electron optics

J. Sánchez
Fisher vector encoding for the classification of natural images

P. Santi
Light sheet fluorescence microscopy

C.J.R. Sheppard, J. Lin and S.S. Kou
The Rayleigh–Sommerfeld diffraction theory

R. Shimizu, T. Ikuta and Y. Takai
Defocus image modulation processing in real time

T. Soma
Focus-deflection systems and their applications

P. Sussner and M.E. Valle
Fuzzy morphological associative memories

J. Valdés
Recent developments concerning the Système International (SI)

G. Wielgoszewski
Scanning thermal microscopy and related techniques

CONTRIBUTORS

Niels de Jonge

INM—Leibniz Institute for New Materials, Innovate Electron Microscopy Group, Campus D2 2, 66123 Saarbrücken, Germany; Department of Molecular Physiology and Biophysics, Vanderbilt University School of Medicine, 2215 Garland Ave, Nashville, TN 37232-0615; Department of Physics, University of Saarland, Campus A5 1, 66123 Saarbrücken, Germany

Andrey I. Denisyuk

St. Petersburg National Research University of Information Technologies, Mechanics, and Optics (ITMO University), 49 Kronverksky, 197101 St. Petersburg, Russia

Jian-Jiun Ding

Graduate Institute of Communication Engineering, National Taiwan University

Filipp E. Komissarenko

St. Petersburg National Research University of Information Technologies, Mechanics, and Optics (ITMO University), 49 Kronverksky, 197101 St. Petersburg, Russia

Alexey V. Krasavin

Department of Physics, King's College London, Strand, London WC2R 2LS, United Kingdom

Ivan S. Mukhin

St. Petersburg National Research University of Information Technologies, Mechanics, and Optics (ITMO University), 49 Kronverksky, 197101 St. Petersburg, Russia; St. Petersburg Academic University—Nanotechnology Research and Education Centre of the Russian Academy of Sciences, 8/3 Khlopina St., 195220 St. Petersburg, Russia

Diana B. Peckys

INM—Leibniz Institute for New Materials, Innovate Electron Microscopy Group, Campus D2 2, 66123 Saarbrücken, Germany

Soo-Chang Pei

Graduate Institute of Communication Engineering, National Taiwan University

Marina Pfaff

INM—Leibniz Institute for New Materials, Innovate Electron Microscopy Group, Campus D2 2, 66123 Saarbrücken, Germany

CHAPTER ONE

Practical Aspects of Transmission Electron Microscopy in Liquid

Niels de Jonge[a,b,c], Marina Pfaff[a] and Diana B. Peckys[a]

[a]INM—Leibniz Institute for New Materials, Campus D2 2, 66123 Saarbrücken, Germany
[b]Department of Molecular Physiology and Biophysics, Vanderbilt University School of Medicine, 2215 Garland Ave, Nashville, TN 37232-0615
[c]Department of Physics, University of Saarland, Campus A5 1, 66123 Saarbrücken, Germany

Contents

Advances in Imaging and Electron Physics, Volume 186
ISSN: 1076-5670
http://dx.doi.org/10.1016/B978-0-12-800264-3.00001-0

1. INTRODUCTION

Electron microscopy traditionally provides subnanometer resolution on solid specimens in vacuum. Because many research questions involve specimens in a liquid environment, it has been a goal to image liquid specimens with electron microscopy ever since the invention of the electron microscope (von Ardenne, 1941; Ruska, 1942; Parsons et al., 1974). Triggered by the availability of electron transparent thin membranes of high stability (Williamson et al., 2003; Thiberge et al., 2004), electron microscopy of liquid specimens has experienced an upsurge of interest in the recent years (de Jonge & Ross, 2011; de Jonge, 2014) with several applications in in biology (de Jonge et al., 2009), material science, and chemistry (Zheng et al., 2009). The advantage of liquid electron microscopy for the investigation of biological systems is that whole cells can be imaged in their native liquid environment (de Jonge et al., 2009; Nishiyama et al., 2010; Peckys & de Jonge, 2014a), while such samples would otherwise have to be dehydrated and embedded in plastic or frozen, and mostly also thin-sectioned (Kourkoutis, Plitzko, & Baumeister, 2012). It is also possible to image protein complexes in liquid (Matricardi, Moretz, & Parsons, 1972; Mirsaidov et al., 2012; Dukes et al., 2014), so that attempts can be made in the near future to study the structure of proteins in their native liquid surroundings. It seems possible to image dynamic processes in certain biological systems (Sugi et al., 1997), but the inevitable severe effect of electron beam irradiation needs to be considered. Most applications of liquid scanning transmission electron microscopy (STEM) are currently found for the imaging of inorganic samples. The movement of nanoparticles in liquid has been studied by various researchers (Zheng et al., 2009; Ring & de Jonge, 2010; Chen & Wen, 2012; Ring & de Jonge, 2012; White et al., 2012; Dukes et al., 2013; Liu et al., 2013). The growth processes of nanoparticles (Evans et al., 2011; Xin & Zheng, 2012; Yuk et al., 2012; Liao & Zheng, 2013; Niu et al., 2013; Nielsen et al., 2014), nanorods (Liao et al., 2012), dendrites (White et al., 2012; Kraus & de Jonge, 2013), and clusters

(Williamson *et al.*, 2003; Radisic *et al.*, 2006; Li *et al.*, 2012) are important topics of study. Chemical reactions such as the formation of hollow particles by the Kirkendall effect (Niu *et al.*, 2013) or corrosion of aluminum films (Chee *et al.*, 2014) have been imaged with liquid (S)TEM. The most complex application is probably the imaging of dynamic processes in functional microdevices (Sacci *et al.*, 2014; Zeng *et al.*, 2014), such as what occurs during the investigation of Li-ion batteries (Holtz *et al.*, 2014; Mehdi *et al.*, 2014; Sacci *et al.*, 2014). By exploiting the interaction of the incident electrons with a liquid precursor, defined patterns can be created on a substrate (Liu *et al.*, 2012; Grogan *et al.*, 2013; den Heijer *et al.*, 2014).

In this chapter, we will discuss the principles of the two most frequently used technical approaches for electron microscopy of liquid specimens: (1) open systems using environmental scanning electron microscopy (ESEM) and environmental scanning electron microscopy (ETEM), and (2) scanning transmission electron microscopy (STEM) of specimens enclosed between SiN windows. We will describe the involved equipment and include detailed design requirements of one particular liquid flow system as an example of the considerations that need to be involved. The chapter also contains many practical details for carrying out such experiments.

2. MOST FREQUENTLY USED SYSTEMS FOR ELECTRON MICROSCOPY OF LIQUID SPECIMENS

Basically, two different approaches exist to study liquid specimens (de Jonge & Ross 2011).

2.1 Open Systems

Open systems expose the liquid to the vacuum (see Figure 1.1). The vacuum level and the temperature are adjusted to the vapor pressure of the liquid to achieve equilibrium between liquid and vapor. The vacuum level in the specimen chamber can be adjusted with pump-limiting apertures in the electron optical column. This apparatus was invented in the 1940s (Ruska 1942) for TEM (Figure 1.1a) and is now mostly used to study specimens in a gaseous environment (Helveg *et al.*, 2004). A pressure of up to 1 bar can be realized in such environmental chambers (Gai, 2002). A variation of this concept is the usage of ionic liquids, which have such a low vapor pressure that the specimens can be directly imaged in a high vacuum (Huang et al., 2010). A similar approach was used for SEM (Danilatos & Robinson,

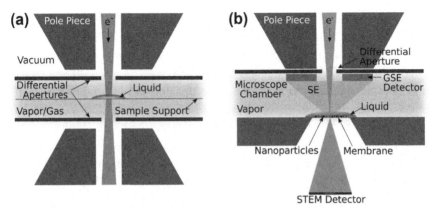

Figure 1.1 Principles for open systems. (a) Schematic drawing of the sample region of ETEM with a differentially pumped environmental chamber containing vapor. (b) The design and working principle of an ESEM with an STEM detector and a gaseous secondary electron detector (GSED). The dimensions are not to scale.

1979) in the 1970s and has found widespread use as variable pressure SEM or ESEM (Stokes, 2003, 2008). A resolution of a few nanometers can be achieved on nanoparticles using the STEM detector (Bogner *et al.*, 2005; Peckys *et al.*, 2013); see Figure 1.1b. The ESEM is now widely used to image specimens in liquid (Peckys *et al.*,, 2013; Barkay, 2014; Bresin *et al.*, 2014; Jansson *et al.*, 2014; Novotny *et al.*, 2014), and is available commercially. In the next sections, we will describe practical aspects of imaging wet samples with ESEM and a STEM detector. Other open systems are not described here for practical reasons.

2.2 Closed Systems

Closed systems protect the liquid from a vacuum by means of thin-membrane windows that are transparent for the electron beam (see Figure 1.2). The crucial aspect of such a system is the material of the membrane. It needs to be made of a light element and as thin as possible to minimize interaction of the electron beam with the membrane, while it must be sufficiently strong to allow practical experiments. Carbon (Daulton *et al.*, 2001; Nishijima *et al.*, 2004) and polymer (Thiberge *et al.*, 2004) foils have been extensively tested, but films of SiN are preferred in practice (Williamson *et al.*, 2003; de Jonge *et al.*, 2009). Delicate sample preparation techniques, developed in recent years, allowed graphene to be used to enclose a small droplet of saline (Mohanty *et al.*, 2011; Yuk *et al.*, 2012). Analytical methods, such as electron energy loss spectroscopy

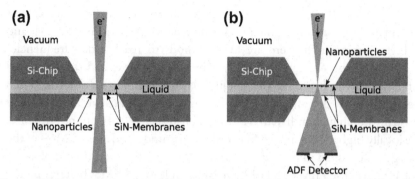

Figure 1.2 Schematic drawing of the sample region of the micro- fluidics system for (a) liquid TEM and (b) liquid STEM. The liquid with nanoparticles is enclosed between two electron-transparent SiN windows forming a channel for liquid flow. The enclosure is placed in the vacuum of the microscope. Images are recorded from the transmitted electron beam with a CCD camera in TEM and with an annular dark-field (ADF) detector in STEM. The dimensions are not to scale.

(Jungjohann *et al.*, 2012) and X-ray dispersive spectrometry, have been demonstrated to function with liquid–cell STEM (Zaluzec *et al.*, 2014). Closed cells are typically used with TEM at 100 keV or more, but closed-cell microchip systems have also been developed for usage in SEM-STEM (Grogan & Bau 2010). In a differently configured SEM experiment, an SiN membrane is used to close the vacuum of the electron optics, whereby the electron beam passes through the membrane onto a specimen in liquid positioned in an open chamber accessible for light microscopy. Backscattered electrons are then used for electron microscopic detection (Nishiyama, Suga *et al.*, 2010). In a different variation, electrons are used to excite a fluorescent layer directly below the sample, thus creating nanoscale light beams (Nawa *et al.*, 2012).

2.3 Choice of Method

The question of whether ESEM (Figure 1.1b), TEM (Figure 1.2a), or STEM (Figure 1.2b) should be used for imaging in liquids is not easy to answer. These techniques have advantages and disadvantages that have to be balanced against each other for each type of experiment. The goal is mostly to achieve a spatial resolution of a few nanometers or more. According to theoretical considerations (de Jonge & Ross, 2011; Schuh & de Jonge, 2014), the resolution of STEM should be better than for TEM, especially for the thicker liquid layers. The resolution of ESEM or SEM of an enclosed liquid is the lowest and typically ranges from 5 nm upward (larger values) for the imaging of nanoparticles in thin liquid layers using ESEM-STEM

(Peckys *et al.*, 2013). If dynamic processes are investigated, the acquisition rate plays a crucial role. Here, TEM usually outperforms STEM and ESEM, even though there are also methods to reach video frequencies in STEM (Ring & de Jonge 2012). The electron dose and the associated influence of the beam on the sample play very important roles in liquid electron microscopy. In conventional TEM, the cumulative electron dose is usually much higher than in STEM (Abellan *et al.*, 2014), but nevertheless, the locally higher dose in STEM can also cause undesired effects. Most of the energy is deposited in the samples using SEM, but the local dose is typically not as high as for TEM or STEM. Last but not least, STEM is better suited to distinguish materials of different atomic numbers (Z) and densities if an annular dark-field (ADF) detector (see Figure 1.2b) is used (Pennycook & Boatner, 1988). The Z-contrast is also highly advantageous when it comes to detecting high-Z nanoparticles in thick liquid layers (de Jonge *et al.*, 2010). In the case of electron microscopy of liquid specimens, the resolution is typically not determined by the intrinsic resolution of the electron micro-scope, but by the sample properties and effects such as beam broadening. To operate liquid TEM or STEM at its limits, plan the experiments, and be able to estimate the achievable resolution, it is important to understand these effects. Details about the expected resolution for these three tech-niques can be found elsewhere (de Jonge & Ross 2011).

3. SYSTEM DESIGN SPECIFICATIONS

3.1 Types of Closed-Cell Systems

The main component of a liquid cell system is a silicon microchip supporting a thin and electron transparent SiN window (Williamson *et al.*, 2003; de Jonge *et al.*, 2007; Zheng *et al.*, 2009; Ring & de Jonge, 2010); see Figure 1.3a. Two microchips are put together with their SiN surfaces facing, so that a thin gap to hold the sample and the liquid is formed between the microchips (see Figure 1.2). In practice, the SiN membranes do not remain flat; they bulge outward into the vacuum on account of the pressure difference between the interior of the liquid cell and the vacuum (see Figure 1.4). Various systems are commercially available or are produced in research labs. Basically, two different concepts can be used. The first is a sealed cell, where the liquid is placed between the windows and the cell is closed thereafter (Williamson *et al.*, 2003; de Jonge *et al.*, 2007, Franks *et al.*, 2008; Liu *et al.*, 2008; Peckys *et al.*, 2009; Zheng *et al.*, 2009; Ring & de Jonge, 2010). Sometimes an additional reservoir is included in the

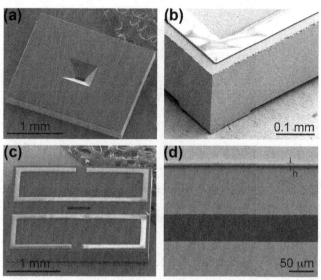

Figure 1.3 SEM images of the microchips. The SEM images were recorded at 10 kV (S4700 Hitachi). (a) Image of the back of a microchip, showing the opening for the SiN window. (b) Close-up of the edge of the microchip, showing the precision- diced edges. The corner of the SU8 spacer can be seen on top of the microchip. (c) Image of the SiN side of the microchip, showing the shape of the SU8 spacer; charging effects distort the image at the positions of the spacer. The SiN window is the dark rectangle in the center. (d) Close-up at the position of the SiN window of the spacer microchip, recorded at a tilt of 45°. The thickness of the spacer layer h was measured to be 6.1 μm. *Image used by permission of* Ring and de Jonge (2010) .

microchip (Williamson *et al.*, 2003; Zheng *et al.*, 2009). A variation of this concept is the enclosure of a droplet between an ultrathin foil (made of a substance such as graphene), wrapping up the liquid sample (Mohanty *et al.*, 2011; Yuk *et al.*, 2012). The second type involves a microfluidic flow system, where the liquid can be continuously replaced and is accessible from the outside of the electron microscope (de Jonge *et al.*, 2009; Ring & de Jonge, 2010). Both systems have their advantages and disadvantages. The thinnest liquid layers can be achieved with the closed–cell approach (Yuk *et al.*, 2012; Dukes *et al.*, 2014). But often, the irradiation of the liquid with the electron beam leads to the creation of bubbles (Peckys *et al.*, 2009; Woehl *et al.*, 2013; Abellan *et al.*, 2014), so that the sample space is not entirely filled with liquid anymore; instead, the specimen on the membrane is imaged in an ultrathin layer of liquid (Figure 1.4) with possibly a high salt concentration, in case the liquid was not pure water. As a result, the achievable resolution in this wet state is good, but after the formation of a

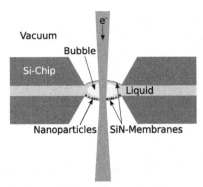

Figure 1.4 Schematic drawing of SiN membranes bulging outward into a vacuum. On account of electron beam irradiation of the specimen, a bubble is formed, so the specimen on the SiN membrane is imaged in only a very thin liquid layer.

bubble, the ion concentration might differ significantly from the applied liquid, and chemistry experiments become difficult to interpret and reproduce. The flow cell principle provides a much better control over the liquid while it is continuously exchanged and thus refreshed. All types of interaction products of the electron beam with the liquid, such as hydrogen, free electrons, radicals, and heat, are carried away from the imaging region. If one needs to image in a liquid layer with a thickness of several hundreds of nanometers to several micrometers, then the only reliable option is to use the flow system.

3.2 Microchips

Figure 1.3a shows the back of a microchip used for liquid TEM/STEM supporting an SiN membrane on the flat side. An SiN membrane of a thickness of 50 nm is typically used, with a liquid thickness of several micrometers (Ring *et al.*, 2011). This thickness balances robustness with sufficient thinness to obtain nanoscale resolution. A rectangular shape of the window of typically $50 \times 400 \ \mu m^2$ can be used. The short dimension is needed to provide stiffness, reducing bulging when the microfluidic chamber is placed in the vacuum of the electron microscope (Figure 1.4). The long dimension is useful in two ways. If two microchips are assembled as a flow cell with parallel windows, the field of view is large, whereas alignment issues are easily resolved when two microchips with elongated windows are assembled perpendicularly (i.e., crossed). A smaller degree of bulging can be obtained by reducing the short dimension of the window. In order to provide easy handling with tweezers and to obtain precise alignment, the edges of the microchips have to be diced with high precision, contrasting a common procedure involving etching and breaking.

To obtain a liquid cell, a second chip with a spacer is used (Figures 1.3b–d). One of the microchips has a spacer (Ring & de Jonge 2010) with a thickness set between 0.5 and 5 μm. Spacers can be made in various ways via lithographic methods or by applying polystyrene microspheres on the edges of a microchip (Ring et al., 2011). Thicker spacers are probably not useful because the resolution becomes too low, while thinner spacers do not make sense because the bulging of the windows and the thickness of the SiN will become limiting factors rather than the spacer. Moreover, in practice, it is not easy to assemble two microchips while maintaining a gap of less than 0.5 μm. If water layers less than 0.5 μm are needed, other configurations (such as pillars) can be used to support the windows (Creemer et al., 2010), a monolithic design (Jensen, Burrows, & Molhave, 2014) or a supported SiN membrane (Dukes et al., 2014). Alternatively, the samples can be wrapped into graphene (Mohanty et al., 2011; Yuk et al., 2012), or electron irradiation can be used to create bubbles in the liquid, reducing the thickness of the liquid (Peckys et al., 2009; Woehl et al., 2013).

For a window with a short dimension of 50 μm and a thickness of 50 nm, the bulging amounts to approximately 1 μm (Figure 1.4). In order to achieve less bulging, chips of a smaller short dimension are available. For the imaging of specimens in water layers less than 0.5 μm thickness, it would make sense to use thinner windows and reduce the window size. Different types of material such as silicon oxide may be tested as well.

3.3 Design Specifications of a Liquid Flow System
3.3.1 Introduction
Various groups manufacture their own microchips and specimen holders, and several companies produce them. It is beyond the scope of this chapter to discuss all possible configurations. As illustration of the considerations involved in a design, we will present two types of microchips used for our research: a base chip and a spacer chip. We will also describe the design specifications for a liquid flow holder fitting these chips.

3.3.2 Specifications of the Base Microchip
A schematic of the base chip is shown in Figure 1.5 with the following specifications (Ring & de Jonge, 2010; Ring et al., 2011): The outer dimensions measure $2.00 \times 2.60 \pm 0.02$ mm. The edges should be precision-made using a technique like precision dicing. If dicing occurs from the top

Topview

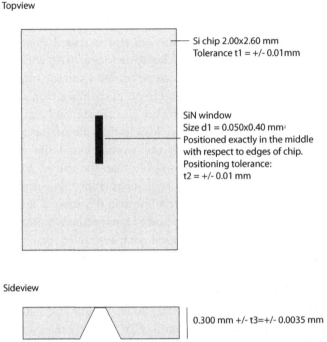

Si chip 2.00x2.60 mm
Tolerance t1 = +/- 0.01mm

SiN window
Size d1 = 0.050x0.40 mm²
Positioned exactly in the middle
with respect to edges of chip.
Positioning tolerance:
t2 = +/- 0.01 mm

Sideview

0.300 mm +/- t3=+/- 0.0035 mm

Figure 1.5 A drawing explaining the specifications of the base chip. The dimensions shown here are not to scale.

(SiN side), the outer dimensions should apply for the bottom plane, such that the chips does not exceed $2.00 \times 2.60 \pm 0.02$ mm over the entire thickness of the chip. The microchip contains an SiN membrane of a typical thickness of 50 nm. The SiN has a highly homogeneous composition and thickness, so the film does not show any features when imaged with an electron beam. It also needs to be sufficiently strong and devoid of holes, so that it can be used to enclose a liquid with a pressure difference of at least 1 bar toward the vacuum. If liquid layers of 0.5 μm and larger are used, the thickness of two SiN membranes of 50 nm each can be ignored, for TEM or STEM imaging at 200 keV, while the resolution is limited by the liquid column and the thickness of one window. But for thinner liquids, or usage with SEM, thinner windows are advantageous.

The chip contains an SiN window. The window is positioned exactly in the middle of the chip, with tolerance t2 = 0.01 mm. Typical dimensions of the windows are as follows:

- V1 = 0.050×0.40 mm^2 (intended for a liquid gap of several micrometers and usage with biological cells)

- V2 = 0.010×0.400 mm^2 (intended for a liquid layer thinner than a micrometer)
- V3 = 0.400×0.010 mm^2 (crossed window for usage with V2 = 0.010×0.40 mm^2, so that an overlapping window can be obtained even if the microchips are not well aligned)

To obtain this high level of precision, use double-sided polished wafers with a precision in thickness of maximal t3 = ± 3.5 μm for a 300-μm wafer, which translates to a tolerance of the window size of $\pm 2 \times 3.5$ μm·tg(54.7°) = ± 10 μm using the KOH etching direction in the silicon of 54.7°.

3.3.3 Spacer Chip

A schematic of the spacer chip (Ring et al., 2011) is shown in Figure 1.6. This chip has the same specifications as for the base chip, but there is an additional spacer layer patterned on the microchip surface. The spacer is defined via photolithography and can be made from SU8 resist material (see Figures 1.3b–d). It has a wall width of d4 = 0.10 mm, with a gap. The spacer should cover a small area but be large enough to provide sufficient robustness and adherence to the microchip. The shape shown in Figure 1.6 provides support over the entire microchip, from the edges to the window, which helps to keep the two microchips as parallel as possible when assembled into a liquid cell. The region at the viewing window is accessible for flowing from the exterior, while the spacer is open at two sides. Each of the two spacers encloses an area on the microchip adjacent to the window. These areas are not completely filled with spacer material for two reasons. First, keeping the spacer area small reduces the chance that unwanted material, such as dust, prevents two microchips from being assembled at the correct distance. Second, it allows excess material to be kept between the chips, thereby reducing the chance that sample material will prevent the microchips from being spaced close together. This is important, for example, when the microchips are loaded with eukaryotic cells growing over the entire surface. Each spacer has a small opening at its outer edge to allow excess material to flow from the enclosed area when the microchips are pressed together. The spacer layer is positioned at a distance of d3 = 0.050 mm from the edge to avoid de-lamination during dicing. The typical window dimensions are the same as for the base chip. The spacer layer thickness usually ranges from 0.5 to 5 μm.

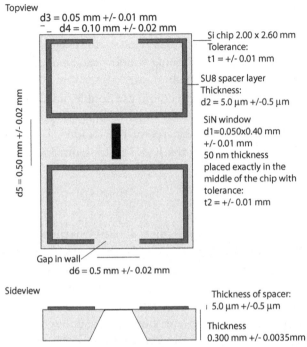

Topview
d3 = 0.05 mm +/- 0.01 mm
d4 = 0.10 mm +/- 0.02 mm

Si chip 2.00 x 2.60 mm
Tolerance:
t1 = +/- 0.01 mm

SU8 spacer layer
Thickness:
d2 = 5.0 μm +/-0.5 μm

SiN window
d1=0.050x0.40 mm
+/- 0.01 mm
50 nm thickness
placed exactly in the
middle of the chip with
tolerance:
t2 = +/- 0.01 mm

d5 = 0.50 mm +/- 0.02 mm

Gap in wall
d6 = 0.5 mm +/- 0.02 mm

Sideview

Thickness of spacer:
5.0 μm +/-0.5 μm

Thickness
0.300 mm +/- 0.0035mm

Figure 1.6 A drawing explaining the specifications of the spacer chip. The dimensions are not to scale.

3.3.4 Liquid Flow Holder

A schematic of a possible liquid flow holder is shown in Figure 1.7. The holder fits two microchips that form a microfluidic channel in a specimen holder for a TEM with a high-resolution pole piece with a typical gap of 5 mm and fitting an objective aperture. The liquid ports in the tip are connected to tubing [such as flexible polyetheretherketone (PEEK) tubing] of an outer diameter of 0.360 mm and an inner diameter of 0.100 mm. Since the tubing can easily become contaminated with sample material, it should be possible for the user to replace it. Ideally, the tip fits in the shaft of the holder and can be replaced as well. The material of the tip and of the lid is a hard metal, biologically inert and nonmagnetic, such as titanium. The chips fit so tightly into a precision-made slot that the windows overlap. The tolerance of the slot is large enough to allow loading of the chips at a slight angle and to allow tolerance of the dimensions of the chips. The extra opening dx (Figure 1.7c) is set at 40 μm, allowing loading at a 20° angle and taking into account the tolerances of the tip and the microchips. The chips are then loaded as straight as possible down in the short dimension. The chips align on the round surfaces of

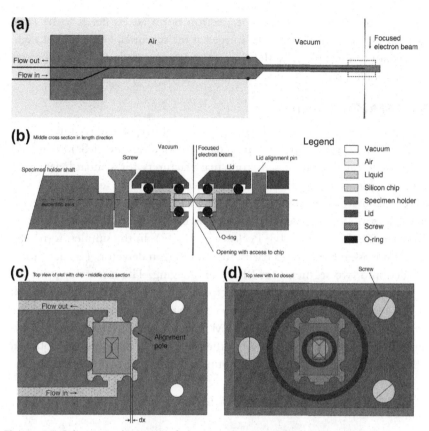

Figure 1.7 Schematic drawings of the tip of a liquid-flow TEM specimen holder. The dimensions are not to scale. (a) The shank of the holder is placed in the vacuum of the electron microscope so that the electron beam passes through the tip. (b) Side view cross section through the tip showing the positioning of the microchips with respect to the electron beam. (c) Top view of the tip with the lid opened so that the liquid path around the microchips. (d) Top view with the lid closed, the color of the lid is transparent so that the positioning of the O-rings can be seen.

alignment poles. The tip closes with a lid, and a configuration with three small O-rings provides vacuum sealing. The lid needs to be able to close within a minute, preferably by three screws. The screws are tightened with a torque screwdriver such that the O-rings are pressurized in a reproducible manner. The microchips thus float on the O-rings, avoiding contact with the sharp protrusions of the metal. Liquid flow through the tip is such that some of it flows between the chips and most of it flows around the chips as a bypass channel. This arrangement is needed so that the tubing can be flushed out in a matter of minutes (Ring & de Jonge 2010). The opening in the lid and the bottom of the holder for electron beam access needs to be large enough so that the sample can be imaged at the specimen holder tilting angle needed for tomography, and the chip can also be accessed with a micropipette tip for washing purposes.

3.4 ESEM Equipment

Liquid experiments with ESEM are carried out by placing samples on a Peltier stage in the microscope chamber (see Figure 1.8a). Standard equipment can be purchased from manufacturers for standard detection using a gaseous secondary electron detector (GSED; see Figure 1.1b). Several laboratories have modified the stage to implement a STEM detector in combination with cooling and those are now also commercially available (Bogner, Thollet et al., 2005, Peckys et al., 2013). In the simplest form, the detector is assembled from a backscattered electron detector. The detector is divided into two segments for BF or DF imaging. The microchip with the sample is then placed above the detector (Figure 1.8b). The main drawback of this configuration is the inability to change the sample independent of the detector. In newer versions, the STEM detector contains multiple segments and can be moved independently of the specimen. The resolution obtained with the STEM detector can be optimized to a few nanometers (Peckys

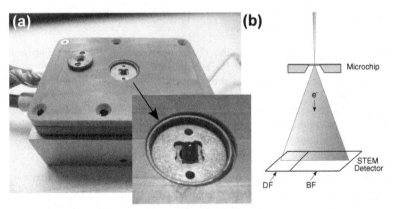

Figure 1.8 ESEM-STEM equipment. (a) Photograph of a Peltier-cooled ESEM stage (Quanta 400 FEG ESEM; FEI, Hillsboro, Oregon). The microchip with an SiN membrane window (shown in the inset at the lower-right corner) resides in a round copper holder such that it is positioned above an integrated STEM detector. Before the closing of the ESEM specimen chamber, the microchip is fixed with the second copper ring, seen on the left side of the stage. (b) Scheme of the integrated STEM detector below the microchip composed of two segments for bright-field (BF) and dark-field (DF) imaging. The dimensions are not to scale. (See the color plate.)

et al., 2013). Since the sample chamber of SEM is typically a few tens of centimeters in size, it is easily possible to integrate additional equipment, such as a light microscope with high numerical aperture (Liv *et al.*, 2013), or a 90° tilting stage (Masenelli-Varlot *et al.*, 2014).

4. PRACTICAL ASPECTS OF ELECTRON MICROSCOPY IN LIQUID

4.1 Liquid-Cell TEM and STEM Experiments

4.1.1 Cleaning the Microchips

To allow high resolution imaging, the SiN window surfaces must be free of any contamination, such as dust particles or material fragments. Both types of contamination are easily collected during manufacturing (especially during dicing), and also during exposure outside a clean-room environment. To keep the microchips clean until they are ready to be used, a resist coating was applied on the SiN during the fabrication process. Once the microchips are ready to be used, the coating is removed by washing for 2 minutes in acetone, followed by a 2-minute wash in ethanol, taking care that the acetone does not dry during the transfer of the microchips between the two liquids. About 50 mL of each fluid are sufficient to clean up to 12 microchips at a time, and high-performance liquid chromatography (HPLC) grade liquids are always used. The microchips should always be checked under a binocular microscope to ensure that the windows are entirely clean (see Figure 1.9). Even the smallest dust particle or piece of broken silicon may prevent the closure of the liquid cell with the desired small spacing.

The windows are hydrophobic after stripping the resistance, but they need to be made hydrophilic for use with most samples by plasma cleaning

Figure 1.9 Photographs showing the difference between coated and clean microchips. (a) A microchip with a protective coating, with embedded debris. (b) The same microchip after washing with acetone, ethanol, and water. *With permission from Ring et al. (2011).*

for several minutes (usually 3–7 minutes). The hydrophilicity lasts for about a day. After plasma cleaning, the microchips are often washed with water (also HPLC grade) to remove any remainders of debris and salt. When the microchips are used with biological cells, they need to be coated with poly-L-lysine (PLL) to extend the duration of their hydrophilic surface and to promote cell adhesion. This coating is applied by immersing the microchips in a solution of 0.01% PLL for 5 minutes at room temperature, followed by two rinses in ultrapure water. The PLL-coated microchips can then be stored in phosphate buffered solution (PBS) until further use; drying should be avoided.

4.1.2 Nonbiological Sample Loading

Samples can be loaded in different ways into the liquid cell, depending on the type of experiment. The most direct approach is to load a droplet with nanoparticles containing liquid and keep this sample in a liquid state until imaging. The nucleation and growth of various types of nanoparticles can be studied this way (Evans et al., 2011). Another approach is to load clean microchips with liquid and then flow the specimen of interest into the viewing area of the microfluidic device, as was demonstrated previously for gold nanoparticles (Ring & de Jonge 2010). For nanoparticle growth experiments, one typically prepares a microchip with seed nanoparticles, dries it, and then applies a droplet of a precursor metal salt solution shortly before imaging (Zheng et al., 2009; Kraus & de Jonge 2013). The correct density of the seed nanoparticles can be checked via conventional (vacuum) TEM or STEM. Similarly, a layer of nanoparticles can be deposited on a microchip and dried (Schuh & de Jonge 2014). Nanoparticles are easily applied from a droplet containing the nanoparticles in a water and ethanol solution, so the droplet quickly dries. The amount of surfactant and salt should be reduced as much as possible to avoid deposits after drying. It is recommended to wash the chip with ethanol and subsequently plasma-clean the chips briefly. The density of nanoparticles can be checked with STEM or TEM. A thin layer of metal can also be vacuum-sputtered onto a membrane, and when the metal layer is sufficiently thin, it breaks up into nanoparticles by itself. Nanoparticles often move or change their shape in liquid when imaged with the electron beam (Ring & de Jonge 2012). To provide stationary samples (for example, to measure the resolution in liquid layers), it is also possible to apply the nanoparticles on the other side of the SiN window; i.e., the side that will be exposed to the vacuum and remains dry (de Jonge et al., 2010). These types of sample preparation result in a high density of nanoparticles on the membrane. Advanced sample geometries are also known in literature, such as microchips containing electrodes and small pieces of battery material (Williamson *et al.*, 2003; Mehdi *et al.*, 2014; Unocic *et al.*, 2014).

4.1.3 Biological Sample Loading

The microchips are compatible for use with cell cultures (de Jonge *et al.*, 2009; Ring *et al.*, 2011). The sample preparation can be done in standard 24- or 96-well plates (see Figure 1.10a). It is recommended to work with 4–10 microchips in parallel. The experiment should be planned in advance such that rows of wells can be prefilled with the liquids needed for the successive processing steps. The actual experiment then merely consists of transferring and incubating microchips from well to well. It is highly recommended to use flat-beak tweezers, preferentially with a soft coating such as polytetrafluorethylene (PTFE; popularly known as Teflon), to avoid flaking of the edges of the microchips. The microchips should remain with their flat side up at all times, and the tweezers should not scratch the surface of the microchip. It is highly advisable to practice the chip-handling beforehand. Details on a typical protocol can be found elsewhere (e.g., Ring *et al.*, 2011; Peckys, Bandmann, & de Jonge 2014).

The microchips with PLL coating are immersed in cell media prior to the seeding of eukaryotic cells. Confluent cells are detached from a culture flask preferably by using cellstripper solution (Mediatech, Manassas, Virginia), and then resuspended in cell culture media. A droplet of the cell suspension is then added to each well containing a microchip and media (Ring *et al.*, 2011). After about 5 minutes, the microchips are inspected with an inverted light microscope to verify the correct density of cells. A microchip should not have more than one adhered cell per 50×50 μm^2 area, as shown in Figure 1.10b. To avoid having too many cells sink down and adhere to the SiN window area, the chips are transferred into new wells prefilled with a cell medium when a sufficient number of cells have attached, usually 7–10 min after the cell seeding started. The chips are then incubated for at 6–24 hours at 37°C in 5% CO_2. Fibroblast cells (such as COS7 cells) adhere to the surface and flatten out, as shown in Figure 1.10c. The cells are then ready for imaging in live state or for further processing.

Nanoparticles of high atomic number materials, such as gold, can be specifically attached to proteins and then be used to detect their locations with STEM (de Jonge *et al.*, 2009; Peckys & de Jonge, 2014a). For example, the

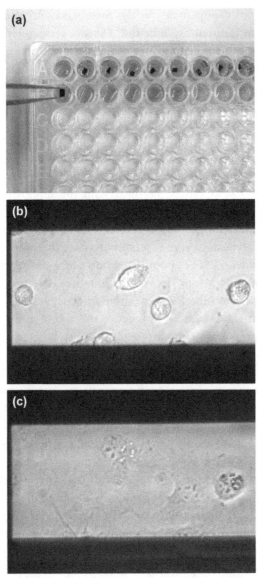

Figure 1.10 Seeding cells onto microchips and labeling the cells. (a) Microchips in a 96-well plate, where they will be seeded with cells and the cells will attach to the surface. Note the orientation of the tweezers; they grip the microchips by their sides, not their top surfaces. Tweezers with Teflon-coated flat tips are recommended. (b) COS7 cells on windows after about 5 minutes of incubation. (c) COS7 cells that have adhered to a window, after about 1 hour of incubation. *With permission from Ring et al. (2011).*

epidermal growth factor receptor (EGFR) can be labeled with gold nanoparticles and studied with STEM in a liquid flow holder (de Jonge et al., 2009), in a closed liquid cell (Peckys et al., 2009), or with ESEM-STEM (Peckys et al., 2013). It is also possible to attach fluorescent quantum dots to the membrane receptors of the cells for use as labels in correlative fluorescence microscopy and liquid STEM (Dukes, Peckys, & de Jonge 2010). Fluorescent labels are useful for inspection with light microscopy, which is recommended after every series of processing steps and can be done directly in the wells. The fixation of cells to SiN windows is needed for electron microscopy, and the process is analogous to the fixation of fluorescence microscopy samples. However, glutaraldehyde is used instead of the typical light microscopy fixatives (i.e., paraformaldehyde or formalin), as it is a stronger fixative (Hopwood 1969). Finally, the microchips are rinsed and stored in a saline buffer solution, such as PBS, at 3 °C.

Smaller cells, such as bacteria and yeast cells, can be pipetted directly onto the microchips prior to imaging (Peckys et al. 2011). Virus particles or protein complexes are typically applied from droplets onto a microchip (Mirsaidov et al., 2012), whereby a coating may promote specific binding of the object of interest (Dukes et al., 2014).

4.1.4 Assembly of the Liquid Cell

The next step in the experiment is to assemble two microchips into a liquid cell, which depends on the type of device. In the first type, the microchips are assembled with liquid between them and are then glued together (Williamson et al., 2003) by placing epoxy at their sides (Peckys et al., 2009; White et al., 2012; Nielsen et al., 2014); see Figure 1.11. The liquid cell is thus closed on all sides, and it can be placed in the vacuum of the electron microscope. The thinnest window is provided in a sample entrapped between graphene sheets (Mohanty et al., 2011; Yuk et al., 2012). Care should be taken that the windows do not break or have holes, so that the sample is actually the liquid one wishes to have, and not a gel containing a high salt concentration obtained after partial drying in the vacuum of the microscope. Alternatively, the liquid cell can be constructed as a monolithic device (Jensen, Burrows, & Molhave, 2014).

In the second type of device, thin windows are placed in a slot and the vacuum sealing is provided by the specimen holder (Parsons, 1974; Sugi et al., 1997), for example, via O-rings (Daulton et al., 2001). This type of device allows for liquid flow between the outside of the microscope and the sample (de Jonge et al., 2009; Ring & de Jonge 2010). The inherent

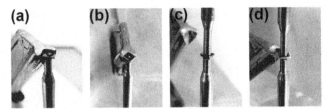

Figure 1.11 Pictures showing assembly of the liquid enclosure. These pictures were made of a test device without a biological sample. (a) The bottom window with a droplet of liquid is positioned on the pole of the loading device with the SiN side facing up. A retractable aligner of the loading device supports two edges. (b) The top window containing the biological specimen is placed face down on the bottom microchip. (c) A pole with a weight presses on the stack of windows. (d) The vacuum epoxy (grey at the sides of the microchips), which serves to glue both microchips together and to vacuum-seal the microchamber, is visible at the sides of the microdevice. *With permission from* Peckys et al. *(2009).*

capability of rapid loading of the sample into the liquid TEM holder allows multiple experiments to be performed per day (Kraus & de Jonge, 2013; Dukes *et al.*, 2014).

Practical aspects of the loading involve the alignment of the microchips and the spacing between the microchips (see Figure 1.12). Ideally, the loading should feature the straightforward placement of the microchips in the holder and the closure of the lid. Regretfully, this is often not easy to accomplish, for several reasons. First, the slot for the placement of the microchips is often not manufactured with sufficient precision and is kept unnecessarily large, so the chips are placed with a tolerance of up to 100 μm in the lateral position and the window of the second microchip does not always overlap that of the first microchip. The loading should thus be done under optical binoculars for precise placement. Second, placing a droplet on a microchip might be difficult because a microchip in a slot that is too large can easily adhere to it. As a result, the droplet cannot be deposited on the microchip; rather, the microchip can cling to the pipette. Finally, the closing procedure can be difficult because it can lead to a bending of the microchips. In order to maintain a liquid gap of 1 μm or less, the microchips must remain absolutely parallel. However, sometimes the microchips are pressed incorrectly into the O–rings, leading to a deformation of the microchips that creates a large gap. After closing, both SiN windows should thus be carefully checked under a binocular microscope for the occurrence of fringe patterns indicating mechanical stress. They should appear entirely flat.

When working with cells or other biological material, the loading procedure is as follows. First, a spacer microchip is positioned with its SiN

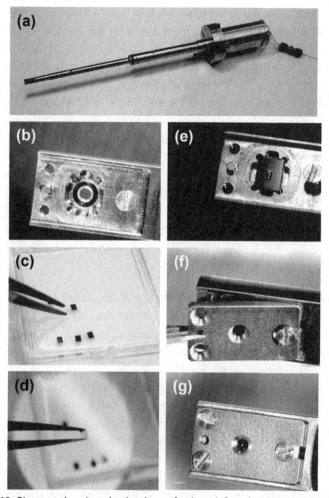

Figure 1.12 Pictures showing the loading of a liquid flow holder. Prototype system made by of Protochips Inc., Raleigh, North Carolina. (a) Picture of the entire holder. (b) Open tip showing the slot with alignment poles and a small O-ring. (c-d) Grabbing a microchip with a flat-beaked tweezer. (e) The slot with the stack of two microchips loaded. (f) Closing the lid. (g) The closed tip.

membrane facing upward, and a droplet of liquid (for example, 10% PBS in water) is pipetted on the window. Glycerol can be added to increase the viscosity and delay evaporation. Preferably within a minute, a microchip with fixed and wet cells is loaded with the cells facing down. The microchip with the cells should be placed gently and directly onto the other chip, while the liquid droplet has not dried out. It is recommended to check the alignment of the windows, but the microchips should not be moved anymore with respect to each other to avoid shearing forces on the sample. The lid needs to be closed immediately, and it should be placed directly in the correct position without causing horizontal movement of the microchips, again to avoid shearing forces. For this reason, the slot should be precision-made, and the alignment of the microchips should be guided (for example, via the alignment poles previously shown in Figure 1.7b). Since the microchip with cells originated from a buffer solution, both sides will be wet. Before the silicon microchip is placed in the electron microscope, the liquid in the window should be removed. Therefore, the backside of the microchip should remain accessible (see Figure 1.7b). To obtain a salt crystal–free outer surface of this membrane, any liquid that remains in the edged window side needs to be replaced with two to three rinses of pure water using a small pipette. The remaining pure water can then be blotted with a small filter paper triangle or simply allowed to evaporate.

4.1.5 Inserting the Liquid Holder into the Microscope

After loading the sample into the liquid holder, the system should be carefully tested for leaking. A practical method of doing this is to initiate a liquid flow through the system (in the case of a liquid flow holder). The holder should be viewed with a binocular microscope to watch for the appearance of liquid droplets on the exterior of the tip or the windows. Again, the windows should appear entirely flat (curved shapes in their corners may indicate the presence of liquid droplets at the outside). Further testing can be done in a dry pump station, but this is typically not needed for a liquid flow system, while it is useful for a closed system.

Prior to inserting the liquid sample holder into the electron microscope, a few tests need to be carried out on the STEM/TEM. First, the microscope has to be aligned on a test sample. Second, the current transmitted through the sample needs to be measured. A simple method is by monitoring the current density using the phosphor screen. This measure is also available in STEM mode. The beam should be entirely captured on the screen. This action depends on the projector lens settings. Third, one needs to keep track of the time that it takes for a pump-down of a dry sample in the load lock, and the vacuum level in the main vacuum chamber of the microscope should also be monitored.

Finally, the holder is ready to be loaded in the TEM/STEM. It is important to monitor the time that it takes to reach the vacuum level in the load lock that is sufficient for transferring the holder into the mean vacuum chamber. If an O-ring leaks, it will be apparent during this stage. As the next test, the holder should be carefully placed in the main vacuum chamber while the vacuum level is monitored. If the vacuum level is less than a factor of 2 of the nominal value, then the sample should be retracted and inspected outside the microscope.

4.1.6 Testing the Correct Liquid Thickness in the Electron Microscope

Once the vacuum level is all right, the gun valve is carefully opened and the specimen is imaged at low magnification (5,000–20,000 x). The first step is to find the lateral (x,y) position of the window at low magnification. Once light comes through, it needs to be checked if the liquid layer is sufficiently thin. This can be done by monitoring the transmitted current. The current is a direct measure of the liquid thickness in case of STEM (de Jonge et al., 2010). The number of electrons N scattered into an angle larger than the opening angle β of the ADF detector can be calculated by the following equation (Reimer & Kohl 2008):

$$N = N_0 \left(1 - \exp\left(-\frac{t}{l}\right) \right) = N_0 \left(1 - \exp\left(-\frac{t\sigma(\beta)N_A}{A} \right) \right), \qquad (1)$$

where N_0 is the number of incoming electrons, t the sample thickness, l the mean free path length, $\sigma(\beta)$ the partial elastic cross section, N_A Avogadro's number, and A the atomic weight. Consequently, the number of electrons detected on the fluorescent screen N_{det} is

$$N_{det} = N_0 - N = N_0 \exp\left(-\frac{t}{l}\right). \qquad (2)$$

If N_{det} and N_0 are measured via the current, the thickness of the sample can easily be calculated by

$$t = -l \ln\left(\frac{N_{det}}{N_0}\right). \qquad (3)$$

For a typical inner dark-field (DF) detector-angle of 70 mrad, the mean free path length l of water is 10.5 μm. Therefore, for a water layer of 5 μm, about 60% of the current hits the screen, while 40% is scattered onto the ADF detector. The exact amount for the specific sample and microscope settings can be easily calculated from the scattering cross sections (de Jonge et al., 2010) or via Monte Carlo calculations (Demers et al., 2010).

TEM applies for liquid layers thinner than a micrometer. An extra test for TEM is to insert an objective aperture (Klein, Anderson, & de Jonge, 2011).

If the liquid layer is too thick, most of the current is then removed as a consequence of most of the electrons being elastically scattered.

If the amount of transmitted current indicates that the liquid is too thick, it might mean that there is a leaking window, so that ice is forming on the vacuum side of the liquid cell. A second possibility is the presence of a liquid with high salt concentration (i.e., gel not evaporating in the vacuum). Another option is that the microchips may be loaded incorrectly so that their SiN windows are spaced too far apart. This occurs, for example, when the liquid cell is assembled from a large and a small microchip and the manufacturing tolerances of the liquid holder are not sufficiently high (Schuh & de Jonge 2014). In this situation, the microchips are incorrectly pressed onto the O-rings, leading to mechanical deformation. We have regularly measured liquid thicknesses of 15, 5 and 5 μm for spacer thicknesses of 5, 2, and 0.5 μm, respectively. These measurements were taken in the middle of the window, where one would expect an increase of the liquid thickness on account of window bending of 2 /character: micro/m maximal. In the corner smaller but still much too large values were measured, for example, 3 /character: micro/m for the 0.5 /character: micro/m spacer. The commercial availability of spacers of specified thicknesses suggests that these liquid gaps can be obtained. But the practically achieved liquid thicknesses often deviate largely from the spacer thickness, even in the corners of the window where the bulging is still small.

4.1.7 Focusing

Once the vacuum is sufficient and the liquid thickness is ascertained, the microscope first should be focused by adjusting the z-position; tilting the stage by 5–10° helps to find the eucentric height. The focus is then adjusted in one of the corners of the window on the edge of the SiN or on small features, such as nanoparticles or small salt or dust deposits. If the windows are very clean and objects are needed for focusing, it is also possible to apply fiducial markers (such as gold nanoparticles on the vacuum side of the top window), whereby the concentration of the liquid from which the nanoparticles are applied has to be adjusted so that salt crusts are avoided and the density of nanoparticles is correct.

The fine focusing and adjusting of the stigmator are sometimes challenging experimental procedures. The SiN window is typically bulged on account of too much pressure inside the liquid cell with respect to the vacuum in the electron microscope (see Figure 1.4). After shifting to a new position with the sample stage, therefore, refocusing is often needed.

On the other hand, the liquid samples are usually prone to changing under electron beam irradiation; e.g., nanoparticles grow as a result of the produced free electrons, and biological cells change their structure on account of radiation damage. As a result, the electron dose needed to focus must be as limited as possible. The focusing procedure can be optimized by means of fiducial markers or other features on the membrane. It might also be possible to map the typical shape of a window beforehand (Holtz *et al.*, 2014), so the focus position can be estimated. The focal depth of the electron beam can be increased by decreasing the beam convergence angle.

4.1.8 Checking for Gas Bubbles

A further complication is the occurrence of gas bubbles. In the case of closed liquid cells, the electron beam creates hydrogen gas bubbles after only a few seconds (Peckys *et al.*, 2009; Woehl *et al.*, 2013; Schneider et al. 2014); see Figure 1.4. The result is a sample configuration containing a thin liquid layer at the membrane and a bubble below that. This situation is not necessarily problematic because the thin liquid layer allows a high spatial resolution (Kraus & de Jonge 2013). However, if the related details, such as window size (critical for the amount of bending), and the liquid thickness, are not published, then it will be difficult for other groups to reproduce the results. One could suspect that many studies that do not include a measure of the liquid thickness, while showing high resolution, might have measured in the presence of bubbles.

The main challenge of electron microscopy in liquid is to extract information of the sample that relates to some of its properties, while the manifold of interactions of the electron beam with the liquid and the solid fractions of the sample causes various additional effects (Grogan et al., 2014). Even if beam-induced effects are used to begin a process, the measurements should be carried out as quantitative as possible so that attempts to understand the measured effects have a chance to succeed. It is, therefore, crucial that researchers describe all relevant sample microscope parameters and include details about how the sample is prepared, how it is loaded into the liquid cell, how to determine the focus, and how to measure the liquid thickness, as well as explicitly mentioning effects such as the occurrence of bubble formation.

4.1.9 Microscopy of Nanomaterials

Liquid TEM and STEM are mostly used for the imaging of growth processes of nanoparticles or nanodendrites in liquid (Zheng *et al.*, 2009; Evans *et al.*, 2011; Li *et al.*, 2012; Liao *et al.*, 2012; Kraus & de Jonge, 2013; Nielsen *et al.*, 2014). TEM is often accomplished in a closed cell, where a bubble formation occurs, so the real liquid layer becomes very thin (Kraus & de Jonge

2013) and the spatial resolution can approach a subnanometer level. When high-Z materials are studied, ADF STEM is advantageous on account of its Z-contrast. Nanomaterials (such as gold nanoparticles of a few nanometers in diameter) are visible with high contrast, even in liquid layers that are several micrometers thick, when imaged with ADF STEM (de Jonge et al., 2010), and atomic resolution is possible for liquid layers that are several hundreds of nanometers thick (de Jonge & Ross, 2011; Jungjohann et al., 2012).

The choice of the thickness of the SiN membrane in an experiment is dictated by the type of sample to be observed. The highest resolution in STEM is achieved for nanoparticles in the top layer of the liquid with respect to a downward-traveling electron beam (see Figure 1.1b). For liquid layers $\geq$ 0.5 μm at 200 keV, the resolution will be noise-limited for most samples and microscope settings (de Jonge et al., 2010; Schuh & de Jonge, 2014), and the influence of 50-nm-thick SiN membranes can be disregarded. For thinner liquid layers, the resolution becomes limited by beam broadening caused by interaction of the electron beam with the SiN membrane and the liquid layer above the focal plane. The material below the focal plane merely contributes to the background noise. An SiN membrane with thickness of 50 nm is recommended to maintain robustness of the liquid cell and to avoid excessive bulging. Atomic resolution can be achieved for nanoparticles in vacuum below an SiN membrane with a thickness of 50 nm (Demers et al., 2012; Ramachandra, Demers, & de Jonge, 2013). It is thus possible to achieve atomic resolution for nanoparticles in liquid adhered to the upper SiN membrane, as was demonstrated in several experiments (e.g., Jungjohann et al., 2012). Using thinner windows for STEM will result in a minor improvement of the spatial resolution and a better signal-to-noise-ratio, but it also causes an increase in practical difficulties. For example, focusing will be more difficult as a consequence of the increased bulging of the thinner window.

TEM needs to be accomplished in a thinner layer of liquid than STEM and, in contrast to STEM, the highest resolution is obtained for nanoparticles at the bottom of the sample (Figure 1.2a). The resolution is typically limited by broadening of the energy spread of the beam in the liquid layer above the sample (inelastic scattering), leading to chromatic aberration (de Jonge & Ross, 2011). The effect of beam broadening in a 50-nm-thick SiN membrane (elastic scattering) can be filtered away using an objective aperture, but at the cost of signal loss. Nevertheless, elastic scattering will likely interfere with phase contrast. The advantage of TEM over STEM in terms of electron dose efficiency will thus diminish when imaging liquid samples.

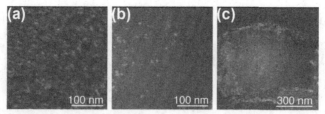

Figure 1.13 Time-resolved STEM imaging of nanoparticles moving in liquid. (a) Image of nanoparticles taken shortly after nanoparticle movement started, recorded at a magnification M = 250,000. The image represents a running average of 8 frames recorded at a video frequency. (b) Image of nanoparticles in the same location, after 22 seconds of exposure to the electron beam. Most of the nanoparticles have left the field of view. (c) Image of the remaining nanoparticles after zooming back out to M = 100,000. Note that nanoparticles that had not been exposed to the electron beam had not been moving. With permission from *Ring and* de Jonge (2012).

Both STEM (Ring & de Jonge, 2012) and TEM (Zheng *et al.*, 2009) can be used to record movies of dynamic processes. An example is shown in Figure 1.13, which depicts a series of STEM images of a layer of gold nanoparticles on an SiN membrane dissolving in the liquid (Ring & de Jonge, 2012; Schuh & de Jonge, 2014). The system shown here was a closed liquid cell. After being imaged at a certain threshold of current density, nanoparticles detached from the SiN membrane and started to move. Most of the nanoparticles in the area irradiated with high current density were eventually removed. Surprisingly, the gold nanoparticles moved three orders of magnitude slower than expected on the base of Brownian motion. Nanoparticles at the SiN membrane and in a thin liquid layer may thus be partially adhered to the surface, and their movement cannot be described by the usual equations for movement in bulk liquid.

4.1.10 Microscopy of Cells

Microscopy of eukaryotic cells can be accomplished using a liquid cell containing a microchip with adhered cells. The typical resolution that can be achieved is a few nanometers for the imaging of gold nanoparticles or quantum dots in the cellular structure with ADF STEM (de Jonge *et al.*, 2009; Dukes *et al.*, 2010). The most difficult experimental procedure is to adjust focus and stigmator settings, so this process should be done in the corner of the window where the liquid is the thinnest. Typically, a first overview image is recorded, as shown in the ADF STEM image of Figure 1.14a for a closed–cell system (Peckys *et al.*, 2009). The sample was loaded as shown in Figure 1.11. If needed, the focus and stigmator

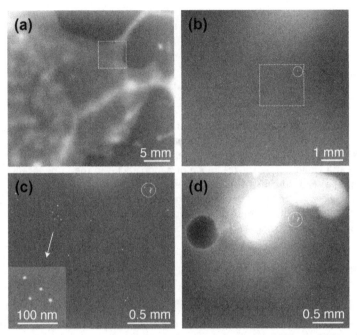

Figure 1.14 STEM imaging of gold-labeled (10-nm diameter) EGFRs in COS7 cells in a wet environment. The beam energy was 200 keV. (a) Image of a region of a cell, showing the cell in lighter gray tones against a darker, uniform background. At this magnification, only clusters of gold labels are visible. (b) Image recorded at the position of the dashed rectangle in (a) at a higher magnification, where labels become visible. (c) The gold labels are visible as individual particles. The inset shows the labels at the highest magnification in this imaging series. (d) Two types of beam damage occurred after the imaging series. A dark round shape is visible at the left, and white shapes are visible at the top.

can be fine-adjusted at an adjacent area at higher magnification than needed for imaging. Once a sharp image is obtained, the stage is shifted to the position of interest, and images are recorded at a magnification between 25,000 and 100,000, which is sufficient to show the nanoparticles. Typically, one zooms in with a few magnifying steps until the structures of interest (for example, nanoparticle labels) become visible, as shown in Figures 1.14b and c. From the locations of the labels, one can derive information about the location of the single proteins and their assembly intoprotein complexes (Peckys *et al.*, 2013, Peckys & de Jonge, 2014a). Radiation damage occurs upon further imaging. Figure 1.14d shows two types of damage. The black shape is a region with reduced scattering as it appears in the ADF detector. Reduced scattering indicates a lower charge density of the material than that of the surrounding material. Together with the round shape, the dark object is interpreted as being a gas bubble. Simultaneously, white shapes appear. These shapes represent deposited contamination of a higher density or larger thickness than the surroundings (de Jonge et al., 2009).

The closed cell provides a good spatial resolution, but the liquid thickness is difficult to control because gas bubbles appear after a while. Furthermore, contamination appears rapidly. These artifacts are avoided with a liquid flow system. A sample with cells can be imaged for an extended period of time, while the cell remains embedded in a known and constant water layer (de Jonge et al., 2009). Radiation damage only appears locally after the recording of several images at a magnification of 100,000, but adjacent areas remain unaffected and can be imaged subsequently.

4.1.11 Correlative Light and Electron Microscopy

It will probably be challenging to record electron microscopic movies of cellular processes on account of radiation damage being many orders of magnitude above the lethal limit for cells (Reimer & Kohl, 2008). The solution when studying biological processes currently seems to be to combine making these movies with light microscopy. Light microscopy can be used to follow certain processes of biological material (for example, via fluorescence microscopy). At a certain point in time, the electron microscope can be used to record high-resolution images of the biological material. The liquid cell can readily be used for correlating light microscopy and electron microscopy. The simplest approach is to image a microchip first with light microscopy, by submerging the microchip in a standard 35 mm plastic dish, filled with PBS. Subsequently, the microchip can be assembled in the microfluidic device, and is then imaged with liquid STEM (Dukes et al., 2010; Ring et al., 2011). The sample (such as one containing live cells) can also be first loaded into the microfluidic device in the tip of the specimen holder. The holder can then be placed on the objective lens of an inverted light microscope (Peckys et al., 2011). By applying a water droplet between the optical lens and the SiN window, light microscopy with high numerical aperture is possible. The sample can be imaged with electron microscopy afterwards. To correlate the light microscopy images with those of the STEM, one simply needs to map the coordinates of the stage to the shape of the SiN window in the light microscopy image, which can be easily done from the corner of the window.

4.2 Liquid ESEM

4.2.1 Introduction

The ESEM can serve as a convenient alternative to liquid cell systems, as it allows for imaging of wet and hydrated samples without the need to protect them from the high-vacuum atmosphere by a special fluid holder. In addition, it avoids having to assemble a fluid cell. Several groups have pioneered the field of biological ESEM imaging (Stokes *et al.*, 2003; Muscariello *et al.*, 2008). The highest resolution can be obtained using the STEM detector, via the so-called wet-STEM technique (Bogner *et al.*, 2005). It is thus possible to image nanoparticle labels of high atomic numbers on hydrated cells with nanometer resolution (Peckys *et al.*, 2013).

4.2.2 Loading the Sample

An experiment typically would be carried out as follows. The stage is first adjusted for the correct sample position, a working distance of typically 6–7 mm, and cooled to 3 °C. The liquid of the specimen should be pure water (or another solvent), and salt should be avoided to prevent the formation of salt crystals. A precooled and wet chip with adherent and labeled cells, rinsed in and covered with pure water, is quickly blotted on the back and placed on the Peltier stage. It is practical to place a few small droplets of pure water (a few microliters each) on the top surface of the stage adjacent to the sample as an additional water source. The low pressure in the microscope specimen chamber is then initiated by a purging sequence, consisting of fivefold pressure cycling between 800 and 1,500 Pa, ending at 800 Pa. Finally, the gun valve is opened and the specimens are examined using a 30-keV electron beam and a GSED.

4.2.3 ESEM Experiments

The thickness of the liquid layer can be adjusted by carefully adjusting the pressure in the column. It is not possible to provide absolute numbers because the pressure depends on several parameters, including the temperature, the amount of liquid, the type of specimen, the region of the sample, and the type of specimen holder or grid. But most cellular samples remain wet for imaging, with pressures ranging between 720 and 750 Pa at 3 °C. It is possible to control the size of water droplets on the sample in experiments (Bresin *et al.*, 2014). As was mentioned previously, the highest resolution can be obtained using the STEM detector on samples containing nanoparticles with a high atomic number. For this purpose, the sample needs to be imaged in transmission mode, and the liquid layer needs to be thinned. A controlled way to reduce the liquid layer while avoiding full drying of the specimen is by simultaneously viewing the GSED and STEM signals while stepwise reducing the pressure starting from 800 Pa to the final imaging pressure. The optimal final pressure lies mostly around 740 Pa, however, sample specific differences in cell morphology and density of the adhering cells can shift the required pressure to slightly higher or lower values. Starting from a thick layer, the signals show background noise only until nanoparticles become visible in the STEM detector after a certain degree of thinning, while the GSED still shows only noise. In this transition point, the atomic number (Z) contrast of the STEM results in a visibility of the high-Z nanoparticles throughout the liquid layer. The nanoparticles also will become visible in the GSED after further thinning of the liquid layer. Typically, focus and stigmator settings are adjusted in a sample region adjacent to the region of interest to avoid radiation damage in the area that is intended to be imaged.

Various sample supports are available for ESEM. ESEM-STEM is often accomplished with standard 3-mm TEM grids, but such grids are difficult to use with cell cultures. We have used silicon microchips supporting SiN windows for the study of whole eukaryotic cells (Ring et al., 2011; Peckys et al., 2013; Peckys & de Jonge, 2014a). Figure 1.15 shows a series of images of whole fixed cells in liquid at increasing magnifications (Peckys et al., 2013). EGFRs in the plasma membrane were labeled with gold nanoparticles. The labels are found as single labels, in pairs (indicating dimerization of the activated receptor), and in higher-order clusters, as shown in Figures 1.15d–g. The ESEM-STEM method was used to study 15 whole cells obtaining data of 1,411 labels (Peckys et al., 2013) The ESEM-STEM method was also used to study the update of gold nanoparticles in A549 human lung carcinoma (Peckys & de Jonge 2014b). Data from 145 cells was collected, resulting in the position data about 1,041 gold nanoparticles. The wet ESEM-STEM method is about a factor of 50 faster for whole-cell studies than for conventional TEM experiments using thin sections. It provides quantitative data on labeled proteins or up-taken nanoparticles from many whole cells.

5. CONCLUSIONS

Electron microscopy of specimens in liquid can be accomplished with three commercially available approaches. The open chamber technique is easily performed using ESEM at a beam energy of several tens of kilovolts and cooled liquid specimens. The highest resolution (to several nanometers) can be achieved for the imaging of high-Z nanoparticles using the ESEM-STEM DF detector. Regular TEM or STEM, using a beam energy of several hundred kilovolts, is mostly accomplished using a liquid enclosure

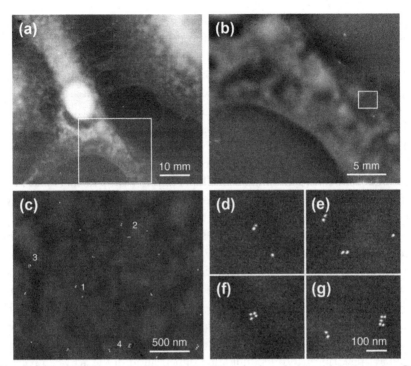

Figure 1.15 SEM of a whole, fixed A549 cell in a hydrated state. (a) An overview DF ESEM-STEM image showing the flat regions of the cells in gray and the thicker cellular areas in white. The pixel size s = 87 nm, and the magnification M = 1,500. (b) Image recorded at the location of the rectangle in (a) using s = 25 nm and M = 5,336. (c) Image showing individual gold nanoparticle labels as white spots for the region shown as a rectangle in (b), s = 2.7 nm and M = 50,000. (d)–(g) Magnified regions from (c), showing individual Au-NPs, dimers, and larger clusters, which are indicated with numbers 1–4 in (c). *With permission from* Peckys et al. (2013).

with thin SiN membranes. Although many groups use TEM, STEM is probably the best for many sample configurations. Atomic resolution can be achieved in liquid enclosures with 50-nm-thick SiN membranes. Dynamic events can be recorded. The experiments demand careful system design, and many factors need to be accounted for during the loading of the specimen and microscopy. Nevertheless, user-friendly systems are now available, and exciting novel science can be expected to result from the new capabilities to image in liquid at nanoscale resolution and beyond.

ACKNOWLEDGMENTS

The microchips and the specimen holder in the photograph were provided by Protochips Inc., based in Raleigh, North Carolina, USA. We thank Marcus Koch for experimental help with the ESEM, and Eduard Arzt for his support through INM. This research was supported in part by the Leibniz Competition 2014.

REFERENCES

Abellan, P., Woehl, T. J., Parent, L. R., Browning, N. D., Evans, J. E., & Arslan, I. (2014). Factors influencing quantitative liquid (scanning) transmission electron microscopy. *Chemical Communications, 50,* 4873–4880.

Barkay, Z. (2014). In situ imaging of nano-droplet condensation and coalescence on thin water films. *Microscopy and Microanalysis, 20,* 317–322.

Bogner, A., Thollet, G., Basset, D., Jouneau, P. H., & Gauthier, C. (2005). Wet STEM: A new development in environmental SEM for imaging nano-objects included in a liquid phase. *Ultramicroscopy, 104,* 290–301.

Bresin, M., Botman, A., Randolph, S. J., Straw, M., & Hastings, J. T. (2014). Liquid phase electron beam–induced deposition on bulk substrates using environmental scanning electron microscopy. *Microscopy and Microanalysis, 20,* 376–384.

Chee, S. W., Duquette, D. J., Ross, F. M., & Hull, R. (2014). Metastable structures in Al thin films before the onset of corrosion pitting as observed using liquid cell transmission electron microscopy. *Microscopy and Microanalysis, 20,* 462–468.

Chen, X., & Wen, J. (2012). In situ wet-cell TEM observation of gold nanoparticle motion in an aqueous solution. *Nanoscale Research Letters, 7,* 598.

Creemer, J. F., Helveg, S., Hoveling, G. H., Ullmann, S., Molenbroek, A. M., Sarro, P. M., & Zandbergen, H. W. (2010). A MEMS reactor for atomic-scale microscopy of nanomaterials under industrially relevant conditions. *Journal of Microelectromechanical Systems, 19,* 254–264.

Danilatos, G. D., & Robinson, V. N. E. (1979). Principles of scanning electron microscopy at high specimen pressures. *Scanning, 18,* 75–78.

Daulton, T. L., Little, B. J., Lowe, K., & Jones-Meehan, J. (2001). In situ environmental cell–transmission electron microscopy study of microbial reduction of chromium(VI) using electron energy loss spectroscopy. *Microscopy and Microanalysis, 7,* 470–485.

de Jonge, N. (2014). Introduction to special issue on electron microscopy of specimens in liquid. *Microscopy and Microanalysis, 20,* 315–316.

de Jonge, N., Peckys, D. B., Kremers, G. J., & Piston, D. W. (2009). Electron microscopy of whole cells in liquid with nanometer resolution. *Proceedings of the National Academy of Sciences, 106,* 2159–2164.

de Jonge, N., Peckys, D. B., Veith, G. M., Mick, S., Pennycook, S. J., & Joy, C. S. (2007). Scanning transmission electron microscopy of samples in liquid (liquid STEM). *Microscopy and Microanalysis, 13*(Suppl. 2), 242–243.

de Jonge, N., Poirier-Demers, N., Demers, H., Peckys, D. B., & Drouin, D. (2010). Nanometer-resolution electron microscopy through micrometer-thick water layers. *Ultramicroscopy, 110,* 1114–1119.

de Jonge, N., & Ross, F. M. (2011). Electron microscopy of specimens in liquid. *Nature Nanotechnology, 6,* 695–704.

Demers, H., Poirier-Demers, N., Drouin, D., & de Jonge, N. (2010). Simulating STEM imaging of nanoparticles in micrometers-thick substrates. *Microscopy and Microanalysis, 16,* 795–804.

Demers, H., Ramachandra, R., Drouin, D., & de Jonge, N. (2012). The probe profile and lateral resolution of scanning transmission electron microscopy of thick specimens. *Microscopy and Microanalysis, 18,* 582–590.

den Heijer, M., Shao, I., Radisic, A., Reuter, M. C., & Ross, F. M. (2014). Patterned electrochemical deposition of copper using an electron beam. *APL Materials, 2,* 022101-1-9.

Dukes, M. J., Jacobs, B. W., Morgan, D. G., Hegde, H., & Kelly, D. F. (2013). Visualizing nano-particle mobility in liquid at atomic resolution. *Chemical Communications, 49*, 3007–3009.

Dukes, M. J., Peckys, D. B., & de Jonge, N. (2010). Correlative fluorescence microscopy and scanning transmission electron microscopy of quantum-dot-labeled proteins in whole cells in liquid. *ACS Nano, 4*, 4110–4116.

Dukes, M. J., Thomas, R., Damiano, J., Klein, K. L., Balasubramaniam, S., Kayandan, S., Riffle, J. S., Davis, R. M., McDonald, S. M., & Kelly, D. F. (2014). Improved microchip design and application for in situ transmission electron microscopy of macromolecules. *Microscopy and Microanalysis, 20*, 338–345.

Evans, J. E., Jungjohann, K. L., Browning, N. D., & Arslan, I. (2011). Controlled growth of nanoparticles from solution with in situ liquid transmission electron microscopy. *Nano Letters, 11*, 2809–2813.

Franks, R., Morefield, S., Wen, J., Liao, D., Alvarado, J., Strano, M., & Marsh, C. (2008). A study of nanomaterial dispersion in solution by wet-cell transmission electron microscopy. *Journal of Nanoscience and Nanotechnology, 8*, 4404–4407.

Gai, P. L. (2002). Development of wet environmental TEM (wet-ETEM) for in situ studies of liquid-catalyst reactions on the nanoscale. *Microscopy and Microanalysis, 8*, 21–28.

Grogan, J. M., & Bau, H. H. (2010). The nanoaquarium: A platform for in situ transmission electron microscopy in liquid media. *Journal of Microelectromechanical Systems, 19*, 885–894.

Grogan, J. M., Schneider, N. M., Ross, F. M., & Bau, H. H. (2013). Bubble and pattern formation in liquid induced by an electron beam. *Nano Letters, 14*, 359–364.

Helveg, S., Lopez-Cartes, C., Sehested, J., Hansen, P. L., Clausen, B. S., Rostrup-Nielsen, J. R., Abild-Pedersen, F., & Norskov, J. K. (2004). Atomic-scale imaging of carbon nanofibre growth. *Nature, 427*, 426.

Holtz, M. E., Yu, Y., Gunceler, D., Gao, J., Sundararaman, R., Schwarz, K. A., Arias, T. A., Abruna, H. D., & Muller, D. A. (2014). Nanoscale imaging of lithium ion distribution during in situ operation of battery electrode and electrolyte. *Nano Letters, 14*, 1453–1459.

Hopwood, D. (1969). A comparison of the crosslinking abilities of glutaraldehyde, formalde-hyde and alpha-hydroxyadipaldehyde with bovine serum albumin and casein. *Histochem-istry and Cell Biology, 17*, 151–161.

Huang, J. Y., Zhong, L., Wang, C. M., Sullivan, J. P., Xu, W., Zhang, L. Q., Mao, S. X., Hudak, N. S., Liu, X. H., Subramanian, A., Fan, H., Qi, L., Kushima, A., & Li, J. (2010). In situ observation of the electrochemical lithiation of a single SnO2 nanowire electrode. *Science, 330*, 1515–1520.

Jansson, A., Boissier, C., Marucci, M., Nicholas, M., Gustafsson, S., Hermansson, A. M., & Olsson, E. (2014). Novel method for visualizing water transport through phase-separated polymer films. *Microscopy and Microanalysis, 20*, 394–406.

Jensen, E., Burrows, A., & Molhave, K. (2014). Monolithic chip system with a microfluidic channel for in situ electron microscopy of liquids. *Microscopy and Microanalysis, 20*, 445–451.

Jungjohann, K. L., Evans, J. E., Aguiar, J. A., Arslan, I., & Browning, N. D. (2012). Atomic-scale imaging and spectroscopy for in situ liquid scanning transmission electron microscopy. *Microscopy and Microanalysis, 18*, 621–627.

Klein, K. L., Anderson, I. M., & de Jonge, N. (2011). Transmission electron microscopy with a liquid flow cell. *Journal of Microscopy, 242*, 117–123.

Kourkoutis, L. F., Plitzko, J. M., & Baumeister, W. (2012). Electron microscopy of biological materials at the nanometer scale. *Annual Review of Materials Research, 42*, 33–58.

Kraus, T., & de Jonge, N. (2013). Dendritic gold nanowire growth observed in liquid with transmission electron microscopy. *Langmuir, 29*, 8427–8432.

Li, D., Nielsen, M. H., Lee, J. R., Frandsen, C., Banfield, J. F., & De Yoreo, J. J. (2012). Direction-specific interactions control crystal growth by oriented attachment. *Science, 336*, 1014–1018.

Liao, H.-G., Cui, L., Whitelam, S., & Zheng, H. (2012). Real-time imaging of Pt3Fe nanorod growth in solution. *Science, 336*, 1011–1014.

Liao, H.-G., & Zheng, H. (2013). Liquid cell transmission electron microscopy study of platinum iron nanocrystal growth and shape evolution. *Journal of the American Chemical Society, 135*, 5038–5043.

Liu, K. L., Wu, C. C., Huang, Y. J., Peng, H. L., Chang, H. Y., Chang, P., Hsu, L., & Yew, T. R. (2008). Novel microchip for in situ TEM imaging of living organisms and bio-reactions in aqueous conditions. *Lab on a Chip, 8*, 1915–1921.

Liu, Y., Chen, X., Noh, K. W., & Dillon, S. J. (2012). Electron beam induced deposition of silicon nanostructures from a liquid phase precursor. *Nanotechnology, 23*, 385302.

Liu, Y., Lin, X.-M., Sun, Y., & Rajh, T. (2013). In situ visualization of self-assembly of charged gold nanoparticles. *Journal of the American Chemical Society, 135*, 3764–3767.

Liv, N., Zonnevylle, A. C., Narvaez, A. C., Effting, A. P., Voorneveld, P. W., Lucas, M. S., Hardwick, J. C., Wepf, R. A., Kruit, P., & Hoogenboom, J. P. (2013). Simultaneous correlative scanning electron and high-NA fluorescence microscopy. *PLoS One, 8*. e55707.

Masenelli-Varlot, K., Malchere, A., Ferreira, J., Heidari Mezerji, H., Bals, S., Messaoudi, C., & Marco Garrido, S. (2014). Wet-STEM tomography: Principles, potentialities, and limitations. *Microscopy and Microanalysis, 20*, 366–375.

Matricardi, V. R., Moretz, R. C., & Parsons, D. F. (1972). Electron diffraction of wet proteins: Catalase. *Science, 177*, 268–270.

Mehdi, B. L., Gu, M., Parent, L. R., Xu, W., Nasybulin, E. N., Chen, X., Unocic, R. R., Xu, P., Welch, D. A., Abellan, P., Zhang, J. G., Liu, J., Wang, C. M., Arslan, I., Evans, J., & Browning, N. D. (2014). In-situ electrochemical transmission electron microscopy for battery research. *Microscopy and Microanalysis, 20*, 484–492.

Mirsaidov, U. M., Zheng, H., Casana, Y., & Matsudaira, P. (2012). Imaging protein structure in water at 2.7 nm resolution by transmission electron microscopy. *Biophysical Journal, 102*, L15–17.

Mohanty, N., Fahrenholtz, M., Nagaraja, A., Boyle, D., & Berry, V. (2011). Impermeable graphenic encasement of bacteria. *Nano Letters, 11*, 1270–1275.

Muscariello, L., Rosso, F., Marino, G., Barbarisi, M., Cafiero, G., & Barbarisi, A. (2008). Cell surface protein detection with immunlogold labelling in ESEM: Optimisation of the method and semi-quantitative analysis. *Journal of Cellular Physiology, 214*, 769–776.

Nawa, Y., Inami, W., Chiba, A., Ono, A., Miyakawa, A., Kawata, Y., Lin, S., & Terakawa, S. (2012). Dynamic and high-resolution live cell imaging by direct electron beam excitation. *Optics Express, 20*, 5629–5635.

Nielsen, M. H., Li, D., Zhang, H., Aloni, S., Han, T. Y.-J., Frandsen, C., Seto, J., Banfield, J. F., Cölfen, H., & De Yoreo, J. J. (2014). Investigating processes of nano-crystal formation and transformation via liquid cell TEM. *Microscopy and Microanalysis, 20*, 425–436.

Nishijima, K., Yamasaki, J., Orihara, H., & Tanaka, N. (2004). Development of microcap-sules for electron microscopy and their application to dynamical observation of liquid crystals in transmission electron microscopy. *Nanotechnology, 15*, S329–S332.

Nishiyama, H., Suga, M., Ogura, T., Maruyama, Y., Koizumi, M., Mio, K., Kitamura, S., & Sato, C. (2010). Atmospheric scanning electron microscope observes cells and tissues in open medium through silicon nitride film. *Journal of Structural Biology, 169*, 438–449.

Niu, K.-Y., Park, J., Zheng, H., & Alivisatos, A. P. (2013). Revealing bismuth oxide hollow nanoparticle formation by the Kirkendall effect. *Nano Letters, 13*, 5715–5719.

Novotny, F., Wandrol, P., Proska, J., & Slouf, M. (2014). In situ WetSTEM observation of gold nanorod self-assembly dynamics in a drying colloidal droplet. *Microscopy and Microanalysis, 20*, 385–393.

Parsons, D. F. (1974). Structure of wet specimens in electron microscopy. *Science, 186,* 407–414.

Parsons, D. F., Matricardi, V. R., Moretz, R. C., & Turner, J. N. (1974). Electron microscopy and diffraction of wet unstained and unfixed biological objects. *Advances in Biological and Medical Physics, 15,* 161–270.

Peckys, D. B., Bandmann, V., & de Jonge, N. (2014). Correlative fluorescence- and scanning transmission electron microscopy of quantum dot labeled proteins on whole cells in liquid. *Methods in Cell Biology.* in press.

Peckys, D. B., Baudoin, J. P., Eder, M., Werner, U., & de Jonge, N. (2013). Epidermal growth factor receptor subunit locations determined in hydrated cells with environmental scanning electron microscopy. *Scientific Reports, 3,* 2626.

Peckys, D. B., & de Jonge, N. (2014a). Liquid scanning transmission electron microscopy: Imaging protein complexes in their native environment in whole eukaryotic cells. *Microscopy and Microanalysis, 20,* 189–198.

Peckys, D. B., & de Jonge, N. (2014b). Gold nanoparticle uptake in whole cells in liquid examined by environmental scanning electron microscopy. *Microscopy and Microanalysis, 20,* 189–197.

Peckys, D. B., Mazur, P., Gould, K. L., & de Jonge, N. (2011). Fully hydrated yeast cells imaged with electron microscopy. *Biophysical Journal, 100,* 2522–2529.

Peckys, D. B., Veith, G. M., Joy, D. C., & de Jonge, N. (2009). Nanoscale imaging of whole cells using a liquid enclosure and a scanning transmission electron microscope. *PLoS One, 4.* e8214.

Pennycook, S. J., & Boatner, L. A. (1988). Chemically sensitive structure-imaging with a scanning transmission electron microscope. *Nature, 336,* 565–567.

Radisic, A., Vereecken, P. M., Hannon, J. B., Searson, P. C., & Ross, F. M. (2006). Quantifying electrochemical nucleation and growth of nanoscale clusters using real-time kinetic data. *Nano Letters, 6,* 238–242.

Ramachandra, R., Demers, H., & de Jonge, N. (2013). The influence of the sample thickness on the lateral and axial resolution of aberration-corrected scanning transmission electron microscopy. *Microscopy and Microanalysis, 19,* 93–101.

Reimer, L., & Kohl, H. (2008). *Transmission Electron Microscopy: Physics of Image Formation.* New York: Springer.

Ring, E. A., & de Jonge, N. (2010). Microfluidic system for transmission electron microscopy. *Microscopy and Microanalysis, 16,* 622–629.

Ring, E. A., & de Jonge, N. (2012). Video-frequency scanning transmission electron microscopy of moving gold nanoparticles in liquid. *Micron, 43,* 1078–1084.

Ring, E. A., Peckys, D. B., Dukes, M. J., Baudoin, J. P., & de Jonge, N. (2011). Silicon nitride windows for electron microscopy of whole cells. *Journal of Microscopy, 243,* 273–283.

Ruska, E. (1942). Beitrag zur übermikroskopischen Abbildungen bei höheren Drucken. *Kolloid Zeitschrift, 100,* 212–219.

Sacci, R. L., Dudney, N. J., More, K. L., Parent, L. R., Arslan, I., Browning, N. D., & Unocic, R. R. (2014). Direct visualization of initial SEI morphology and growth kinetics during lithium deposition by in situ electrochemical transmission electron microscopy. *Chemical Communications, 50,* 2104–2107.

Schneider, N. M., Norton, M. M., Mendel, B. J., Grogan, J. M., Ross, F. M., & Bau, H. H. (2014). Electron–Water Interactions and Implications for Liquid Cell Electron Microscopy. *The Journal of Physical Chemistry C, early online,* http://dx.doi.org/10.1021/jp507400n.

Schuh, T., & de Jonge, N. (2014). Liquid scanning transmission electron microscopy: Nanoscale imaging in micrometers-thick liquids. *Comptes Rendus Physique, 15,* 214–223.

Stokes, D. J. (2003). Recent advances in electron imaging, image interpretation and applications: environmental scanning electron microscopy. *Philosophical Transactions of the Royal Society of London Series A, 361,* 2771–2787.

Stokes, D. J., Rea, S. M., Best, S. M., & Bonfield, W. (2003). Electron microscopy of mammalian cells in the absence of fixing, freezing, dehydration, or specimen coating. *Scanning, 25*, 181–184.

Stokes, D. L. (2008). *Principles and Practice of Variable Pressure/Environmental Scanning Electron Microscopy (VP-SEM)*. Chichester, West Sussex: Wiley.

Sugi, H., Akimoto, T., Sutoh, K., Chaen, S., Oishi, N., & Suzuki, S. (1997). Dynamic electron microscopy of ATP-induced myosin head movement in living muscle filaments. *Proceedings of the National Academy of Sciences, 94*, 4378–4392.

Thiberge, S., Nechushtan, A., Sprinzak, D., Gileadi, O., Behar, V., Zik, O., Chowers, Y., Michaeli, S., Schlessinger, J., & Moses, E. (2004). Scanning electron microscopy of cells and tissues under fully hydrated conditions. *Proceedings of the National Academy of Sciences, 101*, 3346.

Unocic, R. R., Sacci, R. L., Brown, G. M., Veith, G. M., Dudney, N. J., More, K. L., Walden, F. S., 2nd, Gardiner, D. S., Damiano, J., & Nackashi, D. P. (2014). Quantitative electrochemical measurements using in situ ec-S/TEM devices. *Microscopy and Microanalysis, 20*, 452–461.

von Ardenne, M. (1941). Über ein 200 kV-Universal–Elektronenmikroskop mit Objektabschattungsvorrichtung. *Zeitschrift für Physik, 117*, 657–688.

White, E. R., Mecklenburg, M., Shevitski, B., Singer, S. B., & Regan, B. C. (2012). Charged nanoparticle dynamics in water induced by scanning transmission electron microscopy. *Langmuir, 28*, 3695–3698.

White, E. R., Singer, S. B., Augustyn, V., Hubbard, W. A., Mecklenburg, M., Dunn, B., & Regan, B. C. (2012). In situ transmission electron microscopy of lead dendrites and lead ions in aqueous solution. *ACS Nano, 6*, 6308–6317.

Williamson, M. J., Tromp, R. M., Vereecken, P. M., Hull, R., & Ross, F. M. (2003). Dynamic microscopy of nanoscale cluster growth at the solid-liquid interface. *Nature Materials, 2*, 532–536.

Woehl, T. J., Jungjohann, K. L., Evans, J. E., Arslan, I., Ristenpart, W. D., & Browning, N. D. (2013). Experimental procedures to mitigate electron beam–induced artifacts during in situ fluid imaging of nanomaterials. *Ultramicroscopy, 127*, 53–63.

Xin, H. L., & Zheng, H. (2012). In situ observation of oscillatory growth of bismuth nanoparticles. *Nano Letters, 12*, 1470–1474.

Yuk, J. M., Park, J., Ercius, P., Kim, K., Hellebusch, D. J., Crommie, M. F., Lee, J. Y., Zettl, A., & Alivisatos, A. P. (2012). High-resolution EM of colloidal nanocrystal growth using graphene liquid cells. *Science, 336*, 61–64.

Zaluzec, N. J., Burke, M. G., Haigh, S. J., & Kulzick, M. A. (2014). X-ray energy-dispersive spectrometry during in situ liquid cell studies using an analytical electron microscope. *Microscopy and Microanalysis, 20*, 323–329.

Zeng, Z., Liang, W.-I., Liao, H.-G., Xin, H. L., Chu, Y.-H., & Zheng, H. (2014). Visualization of electrode–electrolyte interfaces in LiPF6/EC/DEC electrolyte for lithium ion batteries via in situ TEM. *Nano Letters, 14*, 1745–1750.

Zheng, H., Smith, R. K., Jun, Y. W., Kisielowski, C., Dahmen, U., & Alivisatos, A. P. (2009). Observation of single colloidal platinum nanocrystal growth trajectories. *Science, 324*, 1309–1312.

CHAPTER TWO

Linear Canonical Transform

Jian-Jiun Ding and Soo-Chang Pei

Graduate Institute of Communication Engineering, National Taiwan University

Contents

Advances in Imaging and Electron Physics, Volume 186
ISSN: 1076-5670
http://dx.doi.org/10.1016/B978-0-12-800264-3.00002-2

1. INTRODUCTION

The linear canonical transform (LCT) is a very advanced signal processing tool. It is a generalization of the Fourier transform (FT) and the fractional Fourier transform (FRFT) and can analyze a signal in the domain between time and frequency. Before introducing the LCT, we first review the FT and the FRFT. The FT (Bracewell, 2000) is a very popular operation for spectrum analysis and many other applications. It is defined as

$$\textbf{FT: } F(\omega) = FT(f(t)) = \frac{1}{\sqrt{2\pi}} \int_{-\infty}^{\infty} e^{-j.\omega t} f(t).dt. \tag{1}$$

Its inverse [i.e., the inverse Fourier transform (IFT)] is defined as

$$\textbf{IFT: } F(t) = IFT(f(\omega)) = \frac{1}{\sqrt{2\pi}} \int_{-\infty}^{\infty} e^{-j.\omega t} f(\omega).dt. \tag{2}$$

It is well known that when one performs the FT two times (or $4N+2$ times), the time-reverse operation is obtained. When one performs the FT three times (or $4N+3$ times), the IFT is obtained. Furthermore, performing the FT four times (or $4N$ times) is equivalent to performing an identity operation.

One may ask what will be obtained when the FT is performed a noninteger number of times. The FRFT can be viewed as performing the FT $2\alpha/\pi$ times, where $2\alpha/\pi$ can be a noninteger value. Its definition is

$$\textbf{FRFT: } F_\alpha(u) = O_F^\alpha(f(t))$$

$$= \sqrt{\frac{1 - j \cot \alpha}{2\pi}} e^{\frac{j}{2}\cot \alpha . u^2} \int_{-\infty}^{\infty} e^{-j.\csc \alpha . ut} e^{\frac{j}{2}\cot \alpha . t^2} f(t) dt. \tag{3}$$

When $\alpha \neq N\pi$, N is an integer,

$$F_{2N\pi}(u) = O_F^{2N\pi}(f(t)) = f(u), F_{(2N+1)\pi}(u)$$
$$= O_F^{(2N+1)\pi}(f(t)) = f(-u). \tag{4}$$

It satisfies the additivity property as follows:

$$O_F^{\alpha}\left(O_F^{\beta}(f(t))\right) = O_F^{\beta}\left(O_F^{\alpha}(f(t))\right) = O_F^{\alpha+\beta}(f(t)). \tag{5}$$

It is easy to see that, when $\alpha = 0$, $\pi/2$, π, and $3\pi/2$, the FRFT is reduced to the identity operation, the FT, time-reverse operation, and the IFT, respectively.

The FRFT was first proposed by Wiener (1929) and Condon (1937), and it was reinvented by Namias (1980). Namias derived the FRFT by fractionalizing the eigenvalues of the FT. Then, Almeida (1994), Zayed (1996) and Alieva and Bastiaans (2000), explored the signal processing applications of the FRFT. In Ozaktas, Kutay, and Mendlovic (1999) and Ozaktas, Zalevsky, and Kutay (2000), the theories, properties, and applications of the FRFT were summarized.

Since the FRFT is a generalization of the FT, many properties, applications, and operations associated with FT can be generalized by using the FRFT. The FRFT is more flexible than the FT and performs even better in many signal processing and optical system analysis applications.

However, the FRFT is not general enough since it has only one parameter, α. It can be further generalized into the linear canonical transform (LCT). The LCT has a total of four parameters. It is not only a generalization of the FRFT, but also a generation of the Fresnel transform, the scaling operation.

In the section "Definitions of the Linear Canonical Transform," later in this chapter, we introduce the definitions of the LCT. We also explain the offset LCT, which is a little more general than the LCT and has six parameters. In the next section, "Properties and Physical Meanings of the Linear Canonical Transform," we describe the properties of the LCT, including some basic properties, the transform results for some specific signals, the eigenfunctions and the eigenvalues, the time-frequency properties, the uncertainty principles, and the relations with the random process. Then, in the section "Operations Closely Related to the Linear Canonical Transform," we discuss several operations that are closely related to the LCT, such as the canonical cosine and canonical sine transforms, the simplified FFT, the canonical convolution/correlation operations, and the canonical prolate spheroidal wave function. In the section "Digital Implementation and Discrete Versions of the Linear Canonical Transform," we illustrate how to implement the LCT digitally and introduce several different versions of the discrete LCT. In the next section, "Applications of the Linear Canonical Transform in Signal Processing," we discuss the applications of the LCT in signal processing, including filter design, signal decomposition, signal sampling, signal modulation and multiplexing, acoustics, communication, solving differential equations, encryption, and solving the multi-path problem. In the section "Linear Canonical Transform for Electromagnetic Wave Propagation Analysis," we illustrate how to apply the LCT for electromagnetic wave propagation analysis, such as optics, radar system analysis, and gradient-index medium system analysis. Then, in the section "Two-Dimensional Versions of the Linear Canonical Transform," we introduce and describe two-dimensional versions of the LCT (i.e., the two-dimensional separable and nonseparable linear canonical transform). Finally, we discuss our conclusions.

2. DEFINITIONS OF THE LINEAR CANONICAL TRANSFORM

The LCT was first introduced in the 1970s by Collins (1970) and Moshinsky and Quesne (1971). However, some special cases of LCT with complex parameters were introduced even earlier (e.g., see Bargmann, 1961). In Wolf (1979), a very systematic introduction of the LCT was given. As the FRFT, the LCT was first used for solving differential equations and analyzing optical systems. Recently, after the applications of FRFT were developed, the roles of the LCT for signal processing have also been examined.

The definition of a *linear canonical transform (LCT)* is

$$
\bullet \mathbf{LCT} = F_{(a,b,c,d)}(u) = O_F^{(a,b,c,d)}(f(t))
$$

$$
= \sqrt{\frac{1}{j2\pi b}} . e^{\frac{id}{2b}u^2} \int\limits_{-\infty}^{\infty} e^{-\frac{i}{b}ut} e^{\frac{ia}{2b}t^2} f(t) . dt \quad \text{when } b \neq 0, \tag{6}
$$

$$
F_{(a,0,c,d)}(u) = O_F^{(a,0,c,d)}(f(t)) = \sqrt{d} . e^{\frac{i}{2}cd.u^2} \, f(du) \quad \text{when } b = 0 \tag{7}
$$

The constraint that

$$ad - bc = 1 \qquad (8)$$

should be satisfied. Since there are four parameters $\{a, b, c, d\}$ and one constraint, the degree of freedom of the LCT is 3. The LCT satisfies the additivity property as follows:

$$O_F^{(a_2,b_2,c_2,d_2)} \left[O_F^{(a_1,b_1,c_1,d_1)}(f(t)) \right] = O_F^{(a_3,b_3,c_3,d_3)}(f(t)), \qquad (9)$$

where

$$\begin{bmatrix} a_3 & b_3 \\ c_3 & d_3 \end{bmatrix} = \begin{bmatrix} a_2 & b_2 \\ c_2 & d_2 \end{bmatrix} \cdot \begin{bmatrix} a_1 & b_1 \\ c_1 & d_1 \end{bmatrix}. \qquad (10)$$

In particular,

$$O_F^{(d,-b,-c,a)} \left(O_F^{(a,b,c,d)}(f(t)) \right) = f(t). \qquad (11)$$

That is, the inverse of the LCT with parameters $\{a, b, c, d\}$ is the LCT with parameters $\{d, -b, -c, a\}$. Since the additivity property of LCT can be represented by the multiplication of two 2×2 matrices, as in Eq. (10), we usually use the matrix operation

$$\begin{bmatrix} t \\ f \end{bmatrix} \xrightarrow{LCT} \begin{bmatrix} a & b \\ c & d \end{bmatrix} \cdot \begin{bmatrix} t \\ f \end{bmatrix} \qquad (12)$$

to represent the LCT with parameters $\{a, b, c, d\}$. This is called the *metaplectic representation* (Folland, 1989). For simplification, we call it the *ABCD matrix*.

The LCT has also been called the *special affine Fourier transform (SAFT)* (Abe & Sheridan, 1994b), the *ABCD transform* (Bernardo, 1996), the *generalized Fresnel transform* (James & Agarwal, 1996), the *Collins formula* (Collins, 1970), the *quadratic phase system* (Bastiaans, 1978, 1979, 1989, and 1991), the *generalized Huygens integral* (Siegman, 1986), and the *extended fractional Fourier transform* (Hua, Liu, & Li, 1997).

The LCT is a generalization of many operations, as follows:

(a) When $\{a, b, c, d\} = \{0, 1, -1, 0\}$, the LCT becomes the FT multiplied by $\sqrt{-j}$:

$$O_F^{(0,1-1,0)}(f(t)) = \sqrt{-j}.FT(f(t)). \qquad (13)$$

(b) When $\{a, b, c, d\} = \{0, -1, 1, 0\}$, the LCT becomes the IFT multiplied by $\sqrt{j}$:

$$O_F^{(0,-1,1,0)}(F(\omega)) = \sqrt{j} \cdot IFT(F(\omega)). \qquad (14)$$

(c) When $\{a, b, c, d\} = \{\cos\alpha, \sin\alpha, -\sin\alpha, \cos\alpha\}$, the LCT becomes the FRFT [defined in Eqs. (3) and (4)] multiplied by a constant phase:

$$O_F^{(\cos\alpha,\,\sin\alpha,-\sin\alpha,\,\cos\alpha)}(f(t)) = \left(e^{-j\alpha}\right)^{1/2} O_F^\alpha(F(t)). \qquad (15)$$

(d) When $\{a, b, c, d\} = \{1, \lambda z/2\pi, 0, 1\}$, the LCT becomes the one-dimensional Fresnel transform (i.e., convolution with a chirp). The *Fresnel transform* (Goodman, 2005) describes the propagation of monochromatic light in free space. Suppose that the incoming monochromatic light has a distribution of $U_i(x, y)$, and a wavelength of λ. If it propagates in the free space with length z, then from the Fresnel theory, the output wave has the distribution of $U_o(w, v)$, as follows:

$$U_o(w, v) = \frac{e^{j2\pi \cdot z/\lambda}}{j\lambda z} \int\limits_{-\infty}^{\infty} \int\limits_{-\infty}^{\infty} e^{j\frac{\pi}{\lambda z}(w-x)^2} e^{j\frac{\pi}{\lambda z}(v-y)^2} U_i(x,y).dx\,dy. \qquad (16)$$

It can be rewritten as the combination of two one-dimensional Fresnel transforms:

$$U_o(w, v) = e^{j2\pi \cdot z/\lambda} O_{1D\ Fres}^{y \to v}\left(O_{1D\ Fres}^{x \to w}(U_i(x, y))\right), \qquad (17)$$

where

$$O_{1D\ Fres}^{x \to w} U_i(x, y) = \frac{1}{\sqrt{j\lambda z}} \int\limits_{-\infty}^{\infty} e^{j\frac{\pi}{\lambda z}(w-z)^2} U_i(x, y).dx. \qquad (18)$$

Compared to Eq. (6), one can see that the one-dimensional Fresnel transform corresponds to the LCT with parameters $\{1, \lambda z/2\pi, 0, 1\}$.

(e) When $\{a, b, c, d\} = \{1, 0, \tau, 1\}$, the LCT becomes a chirp multiplication operation:

$$O_F^{(1,0,\tau,1)}(f(t)) = e^{\frac{j}{2}\tau \cdot u^2} f(u). \qquad (19)$$

(f) When $\{a, b, c, d\} = \{\sigma, 0, 0, \sigma^{-1}\}$, the LCT becomes a scaling operation:

$$O_F^{(\sigma,0,0,\sigma^{-1})}(f(t)) = \sqrt{\sigma^{-1}} f(\sigma^{-1}u). \tag{20}$$

Moreover, the LCT can be further generalized. The LCT in Eqs. (6) and (7) has four parameters. In Abe and Sheridan (1994a and 1994b), the LCT can be generalized somewhat by including a space-shifted term and a frequency-modulated term. In this discussion, we call it the *offset LCT*. The offset LCT has six parameters $\{a, b, c, d, m, n\}$.

The definition of the offset LCT is

$$O_F^{(a,b,c,d,m,n)}(f(t)) = \sqrt{\frac{1}{j2\pi b}} e^{jnu} e^{\frac{jd}{2b}(u-m)^2} \int\limits_{-\infty}^{\infty} e^{-\frac{j}{b}(u-m)t} e^{\frac{ja}{2b}t^2} f(t).dt \text{ when } b \neq 0,$$

$$(21)$$

$$O_F^{(a,0,c,d,m,n)}(f(t)) = \sqrt{d} e^{jnu} e^{\frac{j}{2}cd(u-m)^2} \cdot f(d(u-m)) \quad \text{when } b = 0, \tag{22}$$

The constraint that $ad - bc = 1$ must be satisfied. The two extra parameters m and n represent the shifting and the modulation operations, respectively. We can use the following *metaplectic representation* to represent the offset LCT:

$$\begin{bmatrix} u_1 \\ u_2 \end{bmatrix} \xrightarrow{LCT} \begin{bmatrix} a & b \\ c & d \end{bmatrix} \cdot \begin{bmatrix} u_1 \\ u_2 \end{bmatrix} + \begin{bmatrix} m \\ n \end{bmatrix}. \tag{23}$$

For simplification, we call this the *ABCD-MN matrix*. The offset LCT has the additivity property, as follows:

$$O_F^{(a_2,b_2,c_2,d_2,m_2,n_2)}\left[O_F^{(a_1,b_1,c_1,d_1,m_1,n_1)}(f(t))\right] = e^{j\phi} \cdot O_F^{(e,f,g,h,r,s)}(f(t)), \tag{24}$$

where

$$\begin{bmatrix} e & f \\ g & h \end{bmatrix} = \begin{bmatrix} a_2 & b_2 \\ c_2 & d_2 \end{bmatrix} \cdot \begin{bmatrix} a_1 & b_1 \\ c_1 & d_1 \end{bmatrix}, \begin{bmatrix} r \\ s \end{bmatrix} = \begin{bmatrix} a_2 & b_2 \\ c_2 & d_2 \end{bmatrix} \cdot \begin{bmatrix} m_1 \\ n_1 \end{bmatrix} + \begin{bmatrix} m_2 \\ n_2 \end{bmatrix},$$

$$(25)$$

$$\phi = -\frac{a_2c_2}{2}m_1^2 - b_2c_2m_1n_1 - \frac{b_2d_2}{2}n_1^2 - (m_1c_2 + n_1d_2)m_2.$$

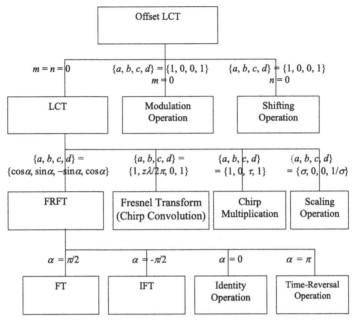

Figure 2.1 Relationships among the offset LCT, the LCT, and their special cases.

The term $\exp(j\phi)$ can be ignored in many conditions since it is independent of t and does not affect the amplitude. The offset LCT is very useful for optical system analysis; for more information, see the section "Linear Canonical Transform for Electromagnetic Wave Propagation Analysis," later in this chapter. The relations among the offset LCT, the LCT, and their special cases are plotted in Fig. 2.1.

3. PROPERTIES AND PHYSICAL MEANINGS OF THE LINEAR CANONICAL TRANSFORM

In this section, the properties of the LCT, including basic properties, the eigenfunctions, the time-frequency properties, the uncertainty principles, and the properties related to random processes, are summarized.

3.1 Basic Properties and Transform Pairs of Linear Canonical Transforms

Table 2.1 shows several transform results of LCTs, and Table 2.2 summarizes some basic properties of the LCT. The properties of the FRFT have been derived systematically by Namias (1980), McBride and Kerr (1987), and

Table 2.1 Transform Results of LCTs for Some Special Signals

Signal	Transform Result (by LCT with parameters $\{a, b, c, d\}$)
$\delta(t-\tau)$	$(j2\pi)^{-\frac{1}{2}} \cdot e^{\frac{jd}{2b}u^2} \cdot e^{\frac{j}{b}u\tau} \cdot e^{\frac{ja}{2b}\tau^2}$
$\exp[-j(qt^2 + rt)]$	$(a - 2qb)^{-\frac{1}{2}} \cdot e^{\frac{jd}{2b}u^2} e^{-j\frac{(u+rb)^2}{2ab-4qb^2}}$
1	$\sqrt{a^{-1}} \cdot e^{j\frac{c}{2a}u^2}$
$\exp(j\tau t)$	$\sqrt{a^{-1}} \cdot e^{j\frac{c}{2a}u^2} \cdot e^{\frac{j\tau}{a}u} \cdot e^{-j\frac{b}{2a}\tau^2}$ $(\text{Im}(\tau) \geq 0)$
$\exp(jht^2/2)$	$(hb + a)^{-\frac{1}{2}} \cdot e^{j\frac{(c+hd)}{2(a+hb)}\cdot u^2}$ $(\text{Im}(h) \geq 0)$
t	$\sqrt{a^{-3}} \cdot u e^{j\frac{c}{2a}u^2}$

Almeida (1994). The properties of the LCT listed in Tables 2.1 and 2.2 can be derived in similar ways.

Most of the transform pairs in Table 2.1 can be proved with the following formula (Spiegel, 2009):

$$\int_{-\infty}^{\infty} e^{-(pt^2+qt)} \cdot dt = \sqrt{\pi/p} \cdot e^{\frac{q^2}{4p}}. \tag{26}$$

Three of the properties in Table 2.2 can be proved as described next.
Proof of the multiplication property:
Differentiating Eq. (6) by u, we obtain

$$\frac{d}{du} F_{(a,b,c,d)}(u) = \sqrt{\frac{1}{j2\pi b}} \cdot e^{\frac{jd}{2b}u^2} \int_{-\infty}^{\infty} \left(\frac{jd}{b}u - \frac{j}{b}t\right) \cdot e^{-\frac{j}{b}ut} e^{\frac{ja}{2b}t^2} \cdot f(t) \cdot dt,$$

$$\frac{d}{du} F_{(a,b,c,d)}(u) = \frac{jd}{b} u \cdot F_{(a,b,c,d)}(u) - \frac{j}{b} \cdot \sqrt{\frac{1}{j2\pi b}} e^{\frac{jd}{2b}u^2} \int_{-\infty}^{\infty} e^{-\frac{j}{b}ut} e^{\frac{ja}{2b}t^2} t f(t) \cdot dt.$$

After multiplying jb, we obtain the multiplication property.
Proof of the differentiation property:
From the reversible property,

$$f(t) = \sqrt{\frac{j}{2\pi b}} \cdot e^{-\frac{ja}{2b}t^2} \int_{-\infty}^{\infty} e^{\frac{j}{b}tu} e^{-\frac{jd}{2b}u^2} \cdot F_{(a,b,c,d)}(u) \cdot du. \tag{27}$$

Table 2.2 Properties of the LCT

Properties	Formulas
Time shift	$O_F^{(a,b,c,d)}(f(t-\tau)) = e^{-j\frac{ac}{2}\tau^2}e^{j.c\tau.u}\cdot F_{(a,b,c,d)}(u-a\tau)$
Modulation	$O_F^{(a,b,c,d)}(e^{j\eta t}f(t)) = e^{-j\frac{bd}{2}\eta^2}e^{jd\eta.u}\cdot F_{(a,b,c,d)}(u-b\eta)$
Time shift and modulation	$O_F^{(a,b,c,d)}(e^{j\eta t}f(t-\tau)) =$ $e^{j\varphi}e^{j.(c\tau+d\eta).u}\cdot F_{(a,b,c,d)}(u-a\tau-b\eta),$ where $\varphi = -(ac/2)\tau^2 - bc\tau\eta - (bd/2)\eta^2$
Scaling	$O_F^{(a,b,c,d)}(\sqrt{\sigma^{-1}}.f(\sigma^{-1}t)) = O_F^{\left(\sigma a,\frac{b}{\sigma},\sigma c,\frac{d}{\sigma}\right)}(f(t))$
Time-reverse	$O_F^{(a,b,c,d)}(f(-t)) = F_{(a,b,c,d)}(-u)$
Even/odd input → even/odd output	If $f(t) = f(-t)$, then $F_{(a,b,c,d)}(u) = F_{(a,b,c,d)}(-u)$ If $f(t) = -f(-t)$, then $F_{(a,b,c,d)}(u) = -F_{(a,b,c,d)}(-u)$
Multiplication	$O_F^{(a,b,c,d)}(t\,f(t)) = \left(b.j\frac{d}{du} + d.u\right)\cdot F_{(a,b,c,d)}(u)$
Differentiation	$O_F^{(a,b,c,d)}(f'(t)) = \left(a\frac{d}{du} - c.ju\right)\cdot F_{(a,b,c,d)}(u)$
Division	$O_F^{(a,b,c,d)}(f(t)/t) = \frac{j}{b}e^{\frac{j}{2}\frac{d}{b}u^2}\int_{-\infty}^{u} e^{-\frac{j}{2}\frac{d}{b}\cdot z^2}F_{(a,b,c,d)}(z)\cdot dz$
Integration	$O_F^{(a,b,c,d)}\left(\int_{-\infty}^{t} f(t')dt'\right) = \frac{e^{\frac{jc}{2a}u^2}}{a}\int_{-\infty}^{u} e^{\frac{jc}{2a}z^2}F_{(a,b,c,d)}(z)dz$ when $a > 0$, $O_F^{(a,b,c,d)}\left(\int_{-\infty}^{t} f(t')dt'\right) = \frac{e^{\frac{jc}{2a}u^2}}{-a}\int_{u}^{\infty} e^{-\frac{jc}{2a}z^2}F_{(a,b,c,d)}(z)dz$ when $a < 0$
Conjugation	$\overline{O_F^{(a,b,c,d)}(f(t))} = O_F^{(a,-b,-c,d)}(\overline{f(t)})$
Direct current value	$F_{(a,b,c,d)}(0) = \int_{-\infty}^{\infty} e^{\frac{ja}{2b}t^2}f(t)dt$
Conservation of energy (Parseval's theorem)	$\int_{-\infty}^{\infty}|f(t)|^2 dt = \int_{-\infty}^{\infty}\left|F_{(a,b,c,d)}(u)\right|^2 du$
Generalized Parseval's theorem	$\int_{-\infty}^{\infty} f(t)\overline{g(t)}dt = \int_{-\infty}^{\infty} F_{(a,b,c,d)}(u)\overline{G_{(a,b,c,d)}(u)}du$

After differentiation,

$$f'(t) = \sqrt{\frac{j}{2\pi b}} \cdot e^{-\frac{ja}{2b}t^2} \int\limits_{-\infty}^{\infty} \left(-\frac{a}{b}jt + \frac{j}{b}u \right) e^{\frac{j}{b}tu} e^{-\frac{jd}{2b}u^2} \cdot F_{(a,b,c,d)}(u) \cdot du,$$

$$f'(t) + \frac{a}{b}jt \cdot f(t) = \frac{j}{b}\sqrt{\frac{j}{2\pi b}} e^{-\frac{ja}{2b}t^2} \int\limits_{-\infty}^{\infty} e^{\frac{j}{b}tu} e^{\frac{jd}{2b}u^2} u \cdot F_{(a,b,c,d,0,0)}(u) \cdot du,$$

$$O_F^{(a,b,c,d)}\left(f'(t) + \frac{a}{b}jt \cdot f(t) \right) = \frac{j}{b} \cdot u \cdot F_{(a,b,c,d)}(u). \qquad (28)$$

Then, after substituting Eq. (27) into Eq. (28), we obtain the differentiation property.

Proof of Parseval's theorem:

$$\int\limits_{-\infty}^{\infty} \left| F_{(a,b,c,d)}(u) \right|^2 du = \int\limits_{-\infty}^{\infty} F_{(a,b,c,d)}(u) \cdot \overline{F_{(a,b,c,d)}(u)} \cdot du$$

$$= \frac{1}{2\pi |b|} \int\limits_{-\infty}^{\infty} \int\limits_{-\infty}^{\infty} \int\limits_{-\infty}^{\infty} e^{\frac{j}{2b}(at^2 - at'^2)} e^{-\frac{ju}{b}(t - t')} f(t)\overline{f(t')} du\, dt\, dt'.$$

$$(29)$$

Then, from Bracewell (2000),

$$\int\limits_{-\infty}^{\infty} e^{-\frac{ju}{b}(t - t')} du = 2\pi |b| \cdot \delta(t - t'), \qquad (30)$$

we obtain

$$\int\limits_{-\infty}^{\infty} \left| F_{(a,b,c,d)}(u) \right|^2 du = \int\limits_{-\infty}^{\infty} \int\limits_{-\infty}^{\infty} e^{\frac{j}{2b}(at^2 - at'^2)} \delta(t - t') f(t)\overline{f(t')} dt\, dt'$$

$$= \int\limits_{-\infty}^{\infty} |f(t)|^2 dt.$$

3.2 Relations with the Wigner Distribution Function

The Wigner distribution function (WDF) is a quadratic time-frequency distribution (Wigner, 1932; Hlawatsch & Boudreaux-Bartels, 1992; Classen & Mecklenbrauker, 1980). It is defined as

$$
W_f(t, \omega) = 1/2\pi \cdot \int_{-\infty}^{\infty} f(t + \tau/2) \cdot f^*(t - \tau/2) \cdot e^{-j\omega\tau} \cdot d\tau
$$

$$
= 1/2\pi \cdot \int_{-\infty}^{\infty} F(\omega + \eta/2) \cdot F^*(\omega - \eta/2) \cdot e^{jt\eta} \cdot d\eta,
$$

(31)

where $F(\omega)$ is the FT of $f(t)$. In several studies (Lohmann, 1993; Mendlovic & Ozaktas, 1993; Ozaktas, 1994; Lohmann & Soffer, 1994), the relation between the FRFT and the WDF was derived. If $W_f(u, v)$ and $W_{F_\alpha}(u, v)$ are the WDFs of $f(t)$ and $F_\alpha(u)$, respectively, then

$$
W_{F_\alpha}(u, v) = W_f(\cos \alpha \cdot u - \sin \alpha \cdot v, \sin \alpha \cdot u + \cos \alpha \cdot v). \tag{32}
$$

That is, the FRFT is equivalent to rotating the WDF of a function clockwise. In particular, the FT (i.e., $\alpha = \pi/2$) is equivalent to rotating the WDF by 90°.

In fact, in addition to the FRFT, the LCT has very close relations with the WDF (Pei & Ding, 2001a). If $W_f(u, v)$ and $W_{F_{(a,b,c,d)}}(u, v)$ are the WDFs of $f(t)$ and $F_{(a,b,c,d)}(u)$, respectively, then

$$
W_{F_{(a,b,c,d)}}(u, v) = W_f(du - bv, -cu + av); \tag{33}
$$

that is,

$$
W_{F_{(a,b,c,d)}}(au + bv, cu + dv) = W_f(u, v). \tag{34}
$$

Another way to express this is to say that the physical meaning of the LCT is to twist the time-frequency distribution of a function. After performing the LCT, the WDF of a function is twisted, but the area is unchanged. In particular, when $\{a, b, c, d\} = \{1, 0, \tau, 1\}$,

$$
W_{F_{(1,0,\tau,1)}}(u, v) = W_f(u, -\tau u + v). \tag{35}
$$

(In this case, the LCT is reduced to a chirp multiplication operation.) When $\{a, b, c, d\} = \{1, \lambda z/2\pi, 0, 1\}$,

$$
W_{F_{1,\lambda z/2\pi,0,1}}(u, v) = W_f\left(u - \frac{\lambda z}{2\pi} v, v\right). \tag{36}
$$

(In this case, the LCT is reduced to a chirp convolution operation; i.e., the Fresnel transform.) That is, chirp multiplication has the effect of shearing the WDF along the ω-axis and shearing the WDF along the t-axis.

Moreover, as in the case of the FRFT (Alieva & Barbe, 1998), the power of the LCT can be viewed as the Radon transform (i.e., projection along a line) of the WDF:

$$\left|F_{(a,b,c,d)}(u)\right|^2 = \int_{-\infty}^{\infty} W_f(d \cdot u - bv, -cu + av) \cdot dv. \tag{37}$$

In addition to the WDF, the LCT also has a close relation with the ambiguity function (AF; Pei & Ding, 2001a). The AF is defined as

$$A_f(\eta, \tau) = 1/2\pi \cdot \int_{-\infty}^{\infty} f(t + \tau/2) \cdot f^*(t - \tau/2) \cdot e^{-j t \eta} \cdot dt. \tag{38}$$

It has the following relation with the WDF:

$$A_f(\eta, \tau) = IFT_{\omega \to \tau}\{FT_{t \to \eta}[W_f(t, \omega)]\}. \tag{39}$$

If $A_f(\eta, \tau)$ and $A_{F_{(a,b,c,d)}}(\eta, \tau)$ are the AFs of $f(t)$ and its LCT, respectively, then

$$A_{F_{(a,b,c,d)}}(\eta, \tau) = A_f(a\eta - c\tau, -b\eta + d\tau). \tag{40}$$

That is, as the WDF case, after performing the LCT, the AF of a function is twisted.

3.3 Eigenfunctions/Eigenvalues of the Linear Canonical Transform

All the eigenfunctions of the original FT can be expressed as a linear combination of Hermite-Gaussian functions, defined as

$$\phi_m(t) = \exp(-t^2/2) H_m(t), \tag{41}$$

where $H_m(t)$ is the Hermite polynomial of order m.

The explicit form of the Hermite polynomial can be seen from Spiegel (2009). $H_0(t) = 1$, $H_1(t) = t$, $H_2(t) = t^2 - 1$, $H_3(t) = t^3 - 3t$, $H_4(t) = t^4 - 6t^2 + 3$, etc. The corresponding eigenvalues for $\phi_m(t)$ is $(-j)^m$:

$$FT(\phi_m(t)) = (-j)^m \phi_m(\omega). \tag{42}$$

The linear combination of the Hermite-Gaussian functions in Eq. (41) is also the eigenfunctions of the FT (Caola, 1991; Cincotti, Gori, & Santarsiero, 1992):

$$FT\left(\sum_{k=0}^{\infty} a_k\phi_{4k+m}(t)\right) = (-j)^m \sum_{k=0}^{\infty} a_k\phi_{4k+m}(\omega). \tag{43}$$

The eigenfunctions/eigenvalues of the FRFT also were discussed in Namias (1980), McBride and Kerr (1987), and Almeida (1994). The Hermite-Gaussian functions $\phi_m(t)$ in Eq. (41) are also the eigenfunctions of the FRFT, but the eigenvalues are changed to $\exp(-j\alpha m)$ as follows:

$$O_F^\alpha(\phi_m(t)) = \exp(-j\,m\alpha)\phi_m(t), \tag{44}$$

where $\phi_m(x)$ is defined as Eq. (41). Moreover, Mendlovic, Ozaktas, and Lohmann (1994a), Alieva (1996), and Alieva and Barbe (1997) found that when $\alpha = 2N\pi/M$, where N and M are integers, the following functions are also eigenfunctions of the FRFT:

$$\Phi(x) = \sum_{p=0}^{\infty} a_p\phi_{pM+q}(x), \tag{45}$$

where $q = 0, 1, \ldots., M-1$, a_p refers to a constant, and the corresponding eigenvalue is $\exp(-j2qN\pi/M)$.

The eigenfunctions of the LCT are no longer Hermite-Gaussian functions. With the method of operation decomposition, as shown by James and Agarwal (1996), the eigenfunctions and eigenvalues of the LCT for the case where $|a + d| < 2$ were derived. Then, Pei and Ding (2002a) derived the eigenfunctions and eigenvalues of the LCT in other cases and found that the eigenfunction/eigenvalue problem of the LCT can be divided into seven cases and that, for different cases, the styles of the eigenfunctions are also different. The eigenfunctions and eigenvalues of the LCT in these cases can be summarized as follows:

Case A ($|a + d| < 2$)

$$\text{Eigenfunctions}: A \exp\left[\frac{-(1 + i\rho)t^2}{2\sigma^2}\right] H_m\left(\frac{t}{\sigma}\right), \tag{46}$$

where $H_m(t)$ is the mth Hermite polynomial, A is any constant, $\sigma^2 = \dfrac{2|b|}{\sqrt{4-(a+b)^2}}$, and $\rho = \dfrac{\text{sgn}(b)(a-d)}{\sqrt{4-(a+d)^2}}$.

$$\text{Eigenvalues}: \quad \lambda = [\exp(-j\alpha)]^{1/2}\exp(-i\alpha m), \tag{47}$$

where $\alpha = \cos^{-1}\left(\frac{a+d}{2}\right) = \sin^{-1}\left[\frac{\text{sgn}(b)}{2}\sqrt{4-(a+d)^2}\right]$.

Case B (a + d = 2, b = 0)

$$\text{Eigenfunctions}: \quad \sum_{n=0}^{\infty} A_n \delta\left(t - \sqrt{4n\pi|c|^{-1}+h}\right)$$
$$+ \sum_{m=0}^{\infty} B_m \delta\left(t + \sqrt{4m\pi|c|^{-1}+h}\right), \tag{48}$$

where $0 \le h < 4\pi/|c|$ and A_n and B_m are free to choose.

$$\text{Eigenvalues}: \quad \lambda = \exp(jch/2).. \tag{49}$$

Case C (a + d = −2, b = 0)
Case C is the same as Case B, except that $A_n = B_n$ for all n or $A_n = -B_n$ for all n, and that the eigenvalues are changed to $\lambda = \pm(-1)^{1/2}\exp(jch/2)$.

Case D (a + d = 2, b ≠ 0)

$$\text{Eigenfunctions}: \quad \exp\left(j\frac{d-a}{4b}t^2\right) \int_{-\infty}^{\infty} \exp\left[j\frac{(t-q)^2}{2\rho}\right]g(q)\,dq, \tag{50}$$

where $g(t) = \sum_{n=0}^{\infty} C_n \exp\left(jt\sqrt{4n\pi|b|^{-1}+h}\right) + \sum_{m=0}^{\infty} D_m \exp$

$\left(-jt\sqrt{4m\pi|b|^{-1}+h}\right), 0 \le h < 4\pi/|b|$, C_n, D_m, and ρ are free to choose.

$$\text{Eigenvalues}: \quad \lambda = \exp(-jbh/2). \tag{51}$$

Case E (a + d= −2, b ≠ 0)
Case E is the same as Case D, except that $C_n = D_n$ for all n or $C_n = -D_n$ for all n, and that the eigenvalues are changed to $\lambda = \pm(-1)^{1/2}\exp(-jbh/2)$.

Case F (a + d > 2)

$$\text{Eigenfunctions}: \exp\left(\frac{j}{2}\rho t^2\right) \int\limits_{-\infty}^{\infty} \exp\left[\frac{j}{2h}(t-q)^2\right] g(q)dq, \qquad (52)$$

where $g(t)$ is a scaling invariant function (i.e., a fractal) that satisfies

$$\sqrt{\sigma}g(\sigma t) = \lambda g(t) \qquad (53)$$

and

$$\rho = \frac{-2sc}{s(d-a)+\sqrt{(a+d)^2-4}}, h = \frac{sb}{\sqrt{(a+d)^2-4}}, \qquad (54)$$

$$\sigma = \left[a+d \pm \sqrt{(a+d)^2-4}\right]\Big/2, s = \text{sgn}(\sigma - \sigma^{-1}).$$

Eigenvalues: λ [defined as in Eq. (53)].

Case G (a + d < −2)

Case G is the same as Case F, except that $g(t) = \pm g(-t)$ should also be satisfied, s in Eq. (54) is changed to $\text{sgn}(\sigma^{-1}-\sigma)$, and the eigenvalues are changed to $\pm(-1)^{1/2}\lambda$.

In summary, when $|a + d| < 2$ (Case A), the eigenfunction of the LCT is the scaling and chirp multiplication of a Hermite-Gaussian function. When $|a + d| = 2$ and $b \neq 0$ (Cases B and C), the eigenfunction of the LCT is the chirp multiplication of a periodic function. When $|a + d| = 2$ and $b = 0$ (Cases D and E), the eigenfunction is an impulse train. When $|a + d| > 2$ (Cases F and G), the eigenfunction is the chirp convolution and chirp multiplication of a scaling invariant function (i.e., a fractal).

Furthermore, a linear combination of the LCT eigenfunctions with the same eigenvalue is also an eigenfunction of the LCT. Pei and Ding (2003a) further extended these results and derived the eigenfunctions and eigenvalues of the offset LCT. That is, if $E(t)$ is an eigenfunction of the LCT with parameters $\{a, b, c, d\}$ and the corresponding eigenvalue is λ, then

$$E_{m,n}(t) = e^{j\frac{cm+(1-a)n}{2-a-d}t} E\left(t - \frac{(1-d)m+bn}{2-a-d}\right) \qquad (55)$$

is the eigenfunction of the offset LCT with parameters $\{a, b, c, d, m, n\}$, and the corresponding eigenvalue is also λ.

The eigenfunctions of the LCT and the offset LCT are very helpful for analyzing the self-imaging elements of the optics and resonance phenomena in electromagnetic wave propagation systems.

3.4 Uncertainty Principles of the Linear Canonical Transform

The well-known uncertainty principle of the FT was proposed by Heisenberg (1927):

$$\Delta_t^2 \Delta_\omega^2 \geq 1/4, \tag{56}$$

where

$$\Delta_\omega^2 = \int_{-\infty}^{\infty} (\omega - \omega_0)^2 |F(\omega)|^2 d\omega \bigg/ \int_{-\infty}^{\infty} |F(\omega)|^2 d\omega \quad \text{and} \tag{57}$$

$$\Delta_t^2 = \int_{-\infty}^{\infty} (t - t_0)^2 |f(t)|^2 dt \bigg/ \int_{-\infty}^{\infty} |f(t)|^2 dt; \tag{58}$$

$F(\omega)$ is the FT of $f(t)$; and

$$\omega_0 = \int_{-\infty}^{\infty} \omega |X(\omega)|^2 d\omega \bigg/ \int_{-\infty}^{\infty} |X(\omega)|^2 d\omega,$$

$$\tag{59}$$

$$t_0 = \int_{-\infty}^{\infty} t |x(t)|^2 dt \bigg/ \int_{-\infty}^{\infty} |x(t)^2| dt.$$

That is, the product of the variances in the time domain and the frequency domain cannot be smaller than $1/4$. In particular, if $f(t)$ is a Gaussian function,

$$f(t) = A \exp\left[-(t - t_0)^2/2 \right], \tag{60}$$

where A and t_0 are arbitrary constants, $\Delta_t^2 \Delta_\omega^2 = 1/4$ is satisfied.

In Shinde and Gadre (2001), the Heisenberg uncertainty principle was generalized into the case of the FRFT. If Δ_t is defined as in Eq. (58) and

$$\Delta_u^2 = \int_{-\infty}^{\infty} (u - u_0)^2 |F_\alpha(u)|^2 du \bigg/ \int_{-\infty}^{\infty} |F_\alpha(u)|^2 du, \tag{61}$$

where $u_0 = \int_{-\infty}^{\infty} u|F_\alpha(u)|^2 du / \int_{-\infty}^{\infty} |F_\alpha(u)|^2 du$, then the following inequality is satisfied:

$$\Delta_t^2 \Delta_u^2 \geq \sin^2 \alpha / 4. \tag{62}$$

Then, as described in Stern (2008), Zhao et al. (2009), and Guanlei, Xiaotong, and Xiaogang (2010), the uncertainty principle of the FRFT was further extended to the LCT. Suppose that Δ_t is defined in Eq. (58) and

$$\Delta_u^2 = \int_{-\infty}^{\infty} (u - u_0)^2 \left|F_{(a,b,c,d)}(u)\right|^2 du \bigg/ \int_{-\infty}^{\infty} \left|F_{(a,b,c,d)}(u)\right|^2 du,$$

where

$$u_0 = \int_{-\infty}^{\infty} u \left|F_{(a,b,c,d)}(u)\right|^2 du \bigg/ \int_{-\infty}^{\infty} \left|F_{(a,b,c,d)}(u)\right|^2 du, \tag{63}$$

then the following inequality is satisfied:

$$\Delta_t^2 \Delta_u^2 \geq b^2 / 4. \tag{64}$$

Furthermore, Stern (2008), Zhao et al. (2009), Sharma and Joshi (2008), and Guanlei, Xiaotong, and Xiaogang (2009a) derived the uncertainty principle of the LCT for the real signal case:

$$\Delta_t^2 \Delta_u^2 \geq b^2 / 4 + \left(a\Delta_t^2\right)^2. \tag{65}$$

Also, Guanlei, Xiaotong, and Xiaogang, (2009a, 2009b, 2009c, and 2009d) also derived the differential form, the Shannon entropy version, the Renyi entropy version, the Housdorff-Young form, the logarithmic form, and the windowed form of the uncertainty principles for the LCT. The uncertainty principles of the LCT, therefore, are helpful for signal analysis and communication.

3.5 Linear Canonical Transform for Random Process

The random process is very closely related to the WDF and the AF. If $g(t)$ is a random process, then, from Martin and Flandrin (1983) and Flandrin (1988),

$$E\left[W_g(t,\omega)\right] = \frac{1}{2\pi} \int_{-\infty}^{\infty} E[g(t + \tau/2)g^*(t - \tau/2)]e^{-j\omega\tau} d\tau$$

$$= \frac{1}{2\pi} \int_{-\infty}^{\infty} R_g(t,\tau)e^{-j\omega\tau} d\tau = \frac{1}{2\pi} S_g(t,\omega) \tag{66}$$

and

$$E\big[A_g(\eta,\tau)\big] = \frac{1}{2\pi} \int\limits_{-\infty}^{\infty} E[g(t+\tau/2)\cdot g^*(t-\tau/2)]\cdot e^{-jt\eta}\cdot dt$$

$$= \frac{1}{2\pi} \int\limits_{-\infty}^{\infty} R_g(t,\tau)e^{-jt\eta}\cdot dt, \tag{67}$$

where $E(\ )$ is the mean and $R_g(t,\tau)$ and $S_g(t,\omega)$ are the auto-correlation and the power spectral density (PSD) of $g(t)$, respectively. In particular, if $g(t)$ is a stationary random process, Eqs. (66) and (67) become

$$E\big[W_g(t,\omega)\big] = S_g(\omega)/2\pi, E\big[A_g(\eta,\tau)\big] = R_g(\tau)\delta(\eta). \tag{68}$$

Since the LCT is equivalent to the twisting operations of the WDF and the AF [see Eqs. (33) and (40)], the LCT also is very closely related to the random process. Pei and Ding (2010) derived these relations. Some of the relations are described as follows:

- **Theorem 1**: If $G_{(a,b,c,d)}(u)$ is the LCT of a stationary random process $g(t)$, then the auto-correlation functions of $G_{(a,b,c,d)}(u)$ and $g(t)$ have the following relation:

$$R_{G_{(a,b,c,d)}}(u,\tau) = \frac{1}{|a|}e^{j\frac{c}{a}u\tau}R_g\Big(\frac{\tau}{a}\Big), \quad \text{when } a\neq 0 \text{ and } b\neq 0. \tag{69}$$

Note that $|R_{G(a,b,c,d)}(u,\tau)|$ is independent of u. Among the four parameters $\{a, b, c, d\}$ of the LCT, a affects the scaling, c affects the phase, and b and d have no effect on $R_{G(a,b,c,d)}(u,\tau)$.

- **Corollary 1**: When $a = 0$, Eq. (69) cannot be applied, but we can prove that the auto-correlation functions of $G_{(a,b,c,d)}(u)$ and the PSD of $g(t)$ have the following relation:

$$R_{G_{(a,b,c,d)}}(u,\tau) = \delta(\tau)S_g\Big(\frac{u}{b}\Big), \quad \text{when } a = 0 \text{ and } b\neq 0. \tag{70}$$

When $b = 0$, neither Eq. (69) nor Eq. (70) can be applied. In this case, the auto-correlation functions of $G_{(a,b,c,d)}(u)$ and $g(t)$ have the following relation:

$$R_{G_{(a,b,c,d)}}(u,\tau) = |d|e^{j\,c\,d\,u\,\tau}R_g(d\tau), \quad \text{when } b = 0 \text{ and } a\neq 0. \tag{71}$$

- **Corollary 2**: For the Fresnel transform, which is a special case of the LCT where $\{a, b, c, d\} = \{1, \lambda z/2\pi, 0, 1\}$, the relation in Eq. (69) becomes

$$R_{G_{(1,\lambda z/2\pi,0,1)}}(u,\tau) = R_g(\tau). \tag{72}$$

That is, after performing the Fresnel transform for a stationary random process, the auto-correlation function remains unchanged. The stationary random process is invariant to the Fresnel transform.

- **Theorem 2**: If $G_{(a,b,c,d)}(u)$ is the LCT of a stationary random process $g(t)$, the PSD of $G_{(a,b,c,d)}(u)$ and $g(t)$ have the following relations:

$$\text{(i) } S_{G_{(a,b,c,d)}}(u, v) = S_g(av - cu), \quad \text{when } a \neq 0 \text{ and } b \neq 0, \tag{73}$$

$$\text{(ii) } S_{G_{(a,b,c,d)}}(u, v) = S_g(u/b), \quad \text{when } a = 0 \text{ and } b \neq 0, \tag{74}$$

$$\text{(iii) } S_{G_{(a,b,c,d)}}(u, v) = S_g(v/d - cu), \quad \text{when } a \neq 0 \text{ and } b = 0. \tag{75}$$

Note that in the case where $a \neq 0$ and $b \neq 0$, the parameters b and d of the LCT do not have any effect on the PSD.

- **Corollary 3**: If $g(t)$ is a stationary random process [i.e., $S_g(t, \omega)$ is independent of t], then, from Eqs. (73)–(75), one can prove that the LCT of $g(t)$ is still a stationary random process if $c = 0$.
- **Corollary 4**: If $g(t)$ is a white noise [i.e., $S_g(t, \omega)$ is a constant], then, from Eqs. (73) – (75), $S_{G_{(a,b,c,d)}}(u, v)$ is also a constant. In other words, the LCT of white noise is still white noise.

4. OPERATIONS CLOSELY RELATED TO THE LINEAR CANONICAL TRANSFORM

In this section, we introduce several operations that are closely related to the LCT, including the canonical convolution; the canonical correlation; the canonical cosine, sine, and Hartley transforms; the canonical prolate spheroidal wave function; and the simplified FFT.

4.1 Canonical Convolution and Canonical Correlation

The original convolution and correlation operations are defined as

$$\text{Convolution}: z_1(t) = \int\limits_{-\infty}^{\infty} x(t - \tau)y(\tau)d\tau. \tag{76}$$

$$\text{Correlation}: z_2(t) = \int\limits_{-\infty}^{\infty} x(t - \tau)y^*(\tau)d\tau. \tag{77}$$

Remember that the convolution and the correlation operations can be implemented by the FT. If Eqs. (1) and (2) are used as the definitions of the FT and the IFT, then

$$z_1(t) = \sqrt{2\pi} IFT[FT(x(t))FT(y(t))] \text{ and} \tag{78}$$

$$z_2(t) = \sqrt{2\pi} IFT\left[FT(x(t))\overline{FT(y(t))}\right], \tag{79}$$

where the upper bar designates the conjugation. Similar to Eqs. (78) and (79), the canonical convolution and the canonical correlation can be defined by replacing the FT by the LCT with parameters $\{a, b, c, d\}$ and replacing the IFT by the LCT with parameters $\{d, -b, -c, a\}$. The definitions are as follows:

- Canonical convolution:

$$z(t) = O_F^{(d,-b,-c,a)}\left[O_F^{(a,b,c,d)}(x(t))O_F^{(a,b,c,d)}(y(t))\right]. \tag{80}$$

It can also be expressed as the following integral form:

$$z(t) = \frac{1}{2\pi|b|\sqrt{d}}e^{-j\frac{ad+1}{2bd}t^2}$$

$$\times \int_{-\infty}^{\infty}\int_{-\infty}^{\infty} e^{j\frac{t(k+s)-ks}{bd}}e^{j\frac{c}{2d}(k^2+s^2)}x(k)y(s)ds\,dk, \quad \text{when } b\neq 0 \text{ and } d\neq 0,$$

$$\tag{81}$$

$$z(t) = \sqrt{\frac{1}{j2\pi b}}\int_{-\infty}^{\infty} e^{j\frac{a}{b}k(k-t)}x(k)y(t-k)dk, \quad \text{when } b\neq 0 \text{ and } d = 0.$$

$$\tag{82}$$

Since the LCT with $b = 0$ is still in the time domain, it is not meaningful to define the canonical convolution with $b = 0$.

- Canonical correlation:

$$z(t) = O_F^{(d,-b,-c,a)}\left(O_F^{(a,b,c,d)}(x(t))\overline{O_F^{(a,b,c,d)}(y(t))}\right). \tag{83}$$

It can be further generalized into the 12-parameter form as follows:

$$O_{corr}^{(a,b,c,d),(e,f,g,h),(m,n,s,v)}(x(t), y(t)) = O_F^{(m,n,s,v)}$$
$$\left(O_F^{(a,b,c,d)}(x(t)) \overline{O_F^{(e,f,g,h)}(y(t))} \right). \tag{84}$$

The canonical correlation is useful for filter design and the canonical correlation is useful for space variant pattern recognition. For more information, see the section "Applications of the Linear Canonical Transform in Signal Processing," later in this chapter.

4.2 Canonical Hilbert Transform

Moreover, the original Hilbert transform is defined as

$$x_H(t) = IFT(FT(x(t))H(\omega)), \tag{85}$$

where

$$H(u) = -j \text{ for } u > 0, H(u) = j \quad \text{for} \quad u < 0, H(0) = 0.$$

It can also be expressed as a convolution of $x(t)$ and the IFT of $H(\omega)$. Since the convolution can be generalized into the canonical convolution, analogous to Eq. (80), the Hilbert transform can also be generalized into the canonical Hilbert transform (Pei & Ding, 2001a):

$$x_H^{(a,b,c,d)}(t) = O_F^{(d,-b,-c,a)} \left[O_F^{(a,b,c,d)}(x(t))H(u) \right]. \tag{86}$$

The canonical Hilbert transform is a further generalization of the fractional Hilbert transform (Lohmann, Mendlovic, & Zalevsky, 1996; Zayed, 1998). It is useful for bandwidth reduction and asymmetric edge detection (Pei & Ding, 2003b, c).

4.3 Canonical Cosine, Sine, and Hartley Transforms

The canonical cosine transform (CCT) and the canonical sine transform (CST) are generalizations of the cosine and sine transforms. The original cosine and sine transforms are defined as

$$G_C(\omega) = \int_0^\infty \cos(\omega t)g(t)dt, \, G_S(\omega) = \int_0^\infty \sin(\omega t)g(t)dt. \tag{87}$$

When $f(t)$ is an even function, one can use the cosine transform instead of the FT for spectrum analysis. Similarly, when $f(t)$ is odd, the sine transform can be used instead of the FT.

As the original cosine and sine transform, the roles of the CCT and the CST are to replace the LCT when the input is an even or an odd function. The CCT and the CST were derived by Pei and Ding (2002b) as described next:

- CCT

$$G_C^{(a,b,c,d)}(u) = O_C^{(a,b,c,d)}(g(t))$$

$$= \sqrt{\frac{2}{j\pi b}} e^{\frac{i}{2}\frac{d}{b}u^2} \int_0^\infty \cos\left[\frac{u}{b}t\right] e^{\frac{i}{2}\frac{a}{b}t^2} g(t) \cdot dt \quad \text{for } b \neq 0, \tag{88}$$

$$G_C^{(a,b,c,d)}(u) = O_C^{(a,b,c,d)}(g(t)) = \sqrt{d} \cdot e^{\frac{i}{2}cdu^2} g(d \cdot u) \text{ for } b = 0. \tag{89}$$

- CST

$$G_S^{(a,b,c,d)}(u) = O_S^{(a,b,c,d)}(g(t))$$

$$= \sqrt{\frac{2}{j\pi b}} e^{\frac{i}{2}\frac{d}{b}u^2} \int_0^\infty -j\sin\left(\frac{u}{b}t\right) e^{\frac{i}{2}\frac{a}{b}t^2} g(t) \cdot dt \quad \text{for } b \neq 0; \tag{90}$$

$$G_S^{(a,b,c,d)}(u) = O_S^{(a,b,c,d)}(g(t)) = \sqrt{d} \cdot e^{\frac{i}{2}cd\,u^2} g(d \cdot u) \quad \text{for } b = 0. \tag{91}$$

When $\{a, b, c, d\} = \{0, 1, -1, 0\}$, the CCT and the CST are reduced to the original cosine and sine transforms multiplied by constants, respectively:

$$G_C^{(0,1,-1,0)}(u) = \sqrt{\frac{2}{j\pi}} G_C(u), \quad G_S^{(0,1,-1,0)}(u) = \sqrt{\frac{j2}{\pi}} G_S(u). \tag{92}$$

Like the LCT, the CCT and the CST have the additivity property:

$$O_C^{(a_2, b_2, c_2, d_2)}\left(O_C^{(a_1, b_1, c_1, d_1)}(g(t))\right) = O_C^{(a_3, b_3, c_3, d_3)}(g(t)), \quad \text{and}$$

$$O_S^{(a_2, b_2, c_2, d_2)}\left(O_S^{(a_1, b_1, c_1, d_1)}(g(t))\right) = O_S^{(a_3, b_3, c_3, d_3)}(g(t)),$$

$$\tag{93}$$

where

$$\begin{bmatrix} a_2 & b_2 \\ c_2 & d_2 \end{bmatrix} \cdot \begin{bmatrix} a_1 & b_1 \\ c_1 & d_1 \end{bmatrix} = \begin{bmatrix} a_3 & b_3 \\ c_3 & d_3 \end{bmatrix}. \tag{94}$$

When dealing with even symmetric or odd symmetric signals, one can use the CCT or the CST instead of the LCT, which reduces the computation complexity.

Moreover, in Pei *et al.* (1998) and Pei and Ding (2002b), the canonical Harley transform was also derived. The original Hartley transform (Bracewell, 1986) is defined as

$$G_H(\omega) = \int_0^{\infty} \operatorname{cas}(\omega t) g(t) dt, \tag{95}$$

where $\operatorname{cas}(x) = \cos(x) + \sin(x)$. Note that the Hartley transform and the FT [using the definition in Eq. (1)] have the following relation:

$$\begin{aligned} G_H(\omega) &= \sqrt{\frac{\pi}{2}} (1+j) [G(\omega) - jG(-\omega)] \\ &= \sqrt{\frac{\pi}{2}} (1+j) \left[O_F^{(0,1,-1,0)}(\omega) \pm O_F^{(0,-1,1,0)} G(\omega) \right]. \end{aligned} \tag{96}$$

Analogous to Eq. (96), and ignoring the difference in the constant, the canonical Hartley transform can be defined as

$$G_H^{(a,b,c,d)}(u) = \frac{1}{2} \left[G_F^{(a,b,c,d)}(u) + G_F^{(-a,-b,-c,-d)}(u) \right] \quad \text{or} \tag{97}$$

$$G_H^{(a,b,c,d)}(u) = \frac{1}{2} \left[G_F^{(a,b,c,d)}(u) - G_F^{(-a,-b,-c,-d)}(u) \right]. \tag{98}$$

It can be proved that that the canonical Hartley transform defined in Eq. (97) or (98) has the following additivity property:

$$O_H^{(a_2,b_2,c_2,d_2)} \left(O_H^{(a_1,b_1,c_1,d_1)}(g(t)) \right) = O_H^{(a_3,b_3,c_3,d_3)}(g(t)), \tag{99}$$

where $\{a_1, b_1, c_1, d_1\}$, $\{a_2, b_2, c_2, d_2\}$, and $\{a_3, b_3, c_3, d_3\}$ satisfy Eq. (92).

4.4 Generalized Prolate Spheroidal Functions for LCTs

Both the FRFT and the LCT have the energy conservation property and satisfy the following equation:

$$\frac{\int_{-\infty}^{\infty} |F(\omega)|^2 d\omega}{\int_{-\infty}^{\infty} |f(t)|^2 dt} = 1, \frac{\int_{-\infty}^{\infty} \left| F_{(a,b,c,d)}(u) \right|^2 du}{\int_{-\infty}^{\infty} |f(t)|^2 dt} = 1. \tag{100}$$

However, when the supports in the time domain and the frequency domain are no longer $(-\infty, \infty)$, and some energy is lost due to the limitations of the aperture and measurement:

$$0 < E_{fi-FT} = \frac{\int_{-\Omega}^{\Omega} |\tilde{F}(\omega)|^2 d\omega}{\int_{-T}^{T} |f(t)|^2 dt} < 1,$$

$$0 < E_{fi-LCT} = \frac{\int_{-\Omega}^{\Omega} \left| \tilde{F}_{(a,b,c,d)}(u) \right|^2 du}{\int_{-T}^{T} |f(t)|^2 dt} < 1, \tag{101}$$

where $\tilde{F}(\omega)$ and $\tilde{F}_{(a,b,c,d)}(u)$ are the finite Fourier transform (fi–FT) and the finite linear canonical transform (fi–LCT) of $f(t)$.

$$\text{fi} - \text{FT: } \tilde{F}(\omega) = (2\pi)^{-1/2}$$

$$\times \int_{-T}^{T} \exp(-j\omega t)f(t)\, dt \qquad \text{space interval: } t \in [-T, T], \tag{102}$$

$$\text{frequency interval: } \omega \in [-\Omega, \Omega].$$

$$\text{fi} - \text{FRFT}: \tilde{F}_{(a,b,c,d)}(u) == \left(\frac{1}{j2\pi b}\right)^{1/2}$$

$$\times \int_{T_1}^{T_2} \exp\left(\frac{j}{2}\frac{d}{b} u^2 - j\frac{u}{b} t + \frac{j}{2}\frac{a}{b} t^2\right) f(t)\, dt \tag{103}$$

$$ab - bc = 1, t \in [T_1, T_2], \quad u \in (\Omega_1, \Omega_2).$$

Slepian and Pollak (1961) and Landau and Pollak (1961, 1962) proposed the prolate spheroidal wave function (PSWF), which satisfies the following equation:

$$\int_{-T}^{T} K_{F,\Omega}(t_1, t)\, \psi_{n,T,\Omega}(t)\, dt = \lambda_{n,T,\Omega} \psi_{n,T,\Omega}(t_1) \quad \text{where } K_{F,\Omega}(t_1, t)$$

$$= \frac{\sin[\Omega(t_1 - t)]}{\pi(t_1 - t)}.$$

(104)

The PSWFs are sorted according to the values of $\lambda_{n,T,\Omega}$:

$$1 > \lambda_{0,T,\Omega} > \lambda_{1,T,\Omega} > \lambda_{2,T,\Omega} > \ldots\ldots\ldots > 0. \left(\text{All of the } \lambda_{n,T,\Omega} \text{ are real}\right).$$

(105)

Of all the functions, $\psi_{0,T,\Omega}(t)$ is the one that can maximize $E_{fi\text{-}FT}$ in Eq. (101). Among all of the functions that are orthogonal to $\psi_{0,T,\Omega}(t)$, $\psi_{1,T,\Omega}(t)$ is the one that can maximize $E_{fi\text{-}FT}$. Of all the functions that are orthogonal to $\psi_{k,T,\Omega}(t)$ ($k = 0, 1, \ldots, n-1$), $\psi_{n,T,\Omega}(t)$ is the one that can maximize $E_{fi\text{-}FT}$.

Pei and Ding (2005) extend the concept of the PSWF to the case of the fi-LCT. We defined the generalized PSWF as follows. If

$$\eta_{n,T,\Omega} \varphi_{n,T,\Omega}(t_1, t) = \int_{-T}^{T} K_{(a,b,c,d),\Omega}(t_1, t)\, \varphi_{n,T,\Omega}(t)\, dt,$$

(106)

where

$$K_{(a,b,c,d),\Omega}(t_1, t) = \frac{\sin[\Omega(t_1 - t)/|b|]}{\pi(t_1 - t)} \exp\left[\frac{j}{2}\frac{a}{b}\left(t^2 - t_1^2\right)\right], t, t_1 \in [-T, T],$$

(107)

then $\varphi_{n,T,\Omega}(t)$ is called the *generalized PSWF*. In fact, the generalized PSWF has the following relation with the PSWF in Eq. (104):

$$\varphi_{n,T,\Omega}(x) = |b|^{-1/2} \exp\left(-\frac{j}{2}\frac{a}{b}x^2\right) \psi_{n,T/|b|,\Omega}\left(|b|^{-1}x\right).$$

(108)

If the generalized PSWF is sorted according to the values of $\eta_{n,T,\Omega}$:

$$1 > \eta_{0,T,\Omega} > \eta_{1,T,\Omega} > \eta_{2,T,\Omega} > \eta_{3,T,\Omega} > \ldots\ldots\ldots > 0,$$

(109)

then, of all the functions, $\varphi_{0,T,\Omega}(t)$ is the one that can maximize $E_{fi\text{-}LCT}$ in Eq. (101). Of all the functions that are orthogonal to $\varphi_{k,T,\Omega}(t)$ ($k = 0, 1, \ldots, n-1$), $\varphi_{n,T,\Omega}(t)$ is the one that can maximize $E_{fi\text{-}LCT}$.

Like the original PSWF, the generalized PSWF is orthogonal:

$$\int_{-T}^{T} \varphi_{n,T,\Omega}(t)\, \varphi_{m,T,\Omega}^{*}(t)\, dt = 0 \quad \text{if } m \neq n. \tag{110}$$

If $f(t)$ is expanded as a linear combination of the generalized PSWF,

$$f(t) = \sum_{n=0}^{\infty} \sigma_n \varphi_{n,T,\Omega}(t) \quad \text{for } t \in [-T, T], \tag{111}$$

then the power preservation ratio $E_{fi\text{-}LCT}$ can be calculated as follows:

$$E_{fi-LCT} = \frac{\int_{-\Omega}^{\Omega} \left| \tilde{F}_{(a,b,c,d)}(u) \right|^2 du}{\int_{-T}^{T} |f(t)|^2 dt} = \frac{\sum_{n=0}^{\infty} |\sigma_n|^2 \eta_{n,T,\Omega}^2}{\sum_{n=0}^{\infty} |\sigma_n|^2 \eta_{n,T,\Omega}}. \tag{112}$$

4.5 Simplified Fractional Fourier Transform

The FRFT has become very popular in recent years. However, in order to be implemented, the FRFT requires one FT, one scaling operation, and two chirp multiplication operations. Pei and Ding (2000b) proposed a way to simplify the FRFT. Remember that the FRFT is a special case of the LCT where $\{a, b, c, d\} = \{\cos \alpha, \sin \alpha, -\sin \alpha, \cos \alpha\}$. Although the LCT has four parameters, observing Eqs. (6) and (8), we can see that the parameter d does not affect the amplitude of $F_{(a,b,c,d)}(u)$, the parameter b affects only the scale of the tem $\exp(-jut/b)$, and the parameter c depends on a, b, and d. Therefore, for most of the signal processing applications, only the value of

$$b/a \tag{113}$$

has a true effect on the performance of the LCT. For the FRFT, the value of $b/a = \cot \alpha$. The LCT (whose value of b/a is also $\cot \alpha$, but the computation loading is less) can be used instead of the FRFT for signal processing applications.

Pei and Ding (2000b) defined the simplified FRFT as an LCT with parameters

$$\begin{bmatrix} a & b \\ c & d \end{bmatrix} = \begin{bmatrix} \cot \alpha & 1 \\ -1 & 0 \end{bmatrix}. \tag{114}$$

Then, from Eq. (6), the formula of the simplified FRFT is

$$F_{(1),\alpha}(u) = O^{\alpha}_{F(1)}(f(t)) = \sqrt{\frac{1}{j2\pi}} \cdot \int_{-\infty}^{\infty} e^{-jut} e^{\frac{j}{2}t^2 \cot \alpha} f(t) \cdot dt. \qquad (115)$$

Its inverse is an LCT with parameters $\{0, -1, 1, \cot \alpha\}$:

$$F(t) = O^{\alpha}_{IF(1)}\left(f_{(1),\alpha}(u)\right) = \sqrt{\frac{1}{j2\pi}} \cdot e^{\frac{j}{2}t^2 \cot \alpha} \int_{-\infty}^{\infty} e^{jut} F_{(1),\alpha}(u) \cdot du. \qquad (116)$$

Note that the simplified FRFT requires one FT and one chirp multiplication operation. Compared to the original FRFT, one chirp multiplication operation and one scaling operation is saved. Moreover, the fractional convolution can be redefined by the simplified FRFT:

$$z(t) = O^{\alpha}_{IF(1)}\left[O^{\alpha}_{F(1)}(x(t)) O^{\alpha}_{F(1)}(y(t))\right]. \qquad (117)$$

This is a special case of the canonical convolution with parameters $\{\cot \alpha, 1, -1, 0\}$. From Eq. (82), $z(t)$ in Eq. (117) can be expressed as

$$z(t) = \sqrt{\frac{1}{j2\pi}} \int_{-\infty}^{\infty} e^{jk(k-t)\cot \alpha} x(k) y(t-k) dk. \qquad (118)$$

Note that there is only one integral in Eq. (118). By contrast, the fractional convolution defined from the original FRFT requires three integrals (Almeida, 1997). Therefore, the simplified FRFT can also reduce the complexities of computing the fractional convolution and improve the efficiency of the fractional filter.

The simplified FRFT in Eq. (115) is efficient for digital implementation. Pei and Ding (2000b) also defined an alternative version of the simplified FRFT. This is a special case of the LCT, with the following parameters:

$$\begin{bmatrix} a & b \\ c & d \end{bmatrix} = \begin{bmatrix} 1 & \tan \alpha \\ -2 \cot \alpha & -1 \end{bmatrix}. \qquad (119)$$

Its formula is

$$O^{\alpha}_{F(2)}(f(t)) = \sqrt{\frac{\cot \alpha}{j2\pi}} \exp\left(-\frac{j}{2}u^2 \cot \alpha\right)$$

$$\times \int_{-\infty}^{\infty} \exp\left(-jut \cot \alpha + \frac{j}{2}t^2 \cot \alpha\right) f(t) dt. \qquad (120)$$

In optical implementation, when using the original FRFT, two lenses and one free space (or one lens and two free spaces) are required, as discussed in the section "Linear Canonical Transform for Electromagnetic Wave Propagation Analysis," later in this chapter. When using the simplified FRFT in Eq. (120), only one lens and one free space are required for implementation.

5. DIGITAL IMPLEMENTATION AND DISCRETE VERSIONS OF THE LINEAR CANONICAL TRANSFORM

In this section, we introduce the implementation algorithms of the LCT and discuss the proper way to define the discrete version of the LCT.

The conventional FT is popular for two reasons. The first is it has a clear physical meaning, and the second is its fast algorithm. With the fast FT algorithm (Oppenheim & Schafer, 2010), the complexity of the FT is

$$O(N . \log_2 N), \tag{121}$$

where N is the number of sampling points. Similarly, LCTs also have fast algorithms. With these algorithms, the complexity of the LCT is the same as that of the conventional FT. We now discuss two methods of implementing the LCT.

5.1 Chirp Convolution Implementation Method

Before implementing the LCT digitally, first sample the t-axis and u-axis as $p\Delta_t$ and $q\Delta_u$, respectively, and convert Eq. (6) into

$$F_{(a,b,c,d)}(q\Delta_u) = \sqrt{\frac{1}{j2\pi b}} \cdot e^{\frac{j}{2}\frac{d}{b}q^2\Delta_u^2} \sum_{p=-M}^{M} e^{-\frac{j}{b}p \cdot q \cdot \Delta_u\Delta_t} e^{\frac{j}{2}\frac{a}{b}p^2\Delta_t^2} f(p\Delta_t). \tag{122}$$

Then, from Ozaktas and Arikan (1996) and Deng et al. (2000),

$$F_{(a,b,c,d)}(q\Delta_u) = \sqrt{\frac{1}{j2\pi b}} \cdot e^{\frac{j}{2}\frac{d-1}{b}q^2\Delta_u^2}$$

$$\times \sum_{p=-M}^{M} e^{\frac{j}{b}(q\Delta_u - p\Delta_t)^2} e^{\frac{j}{2}\frac{a-1}{b}p^2\Delta_t^2} f(p\Delta_t). \tag{123}$$

Therefore, we can implement the LCT by two chirp multiplications and one chirp convolution:

$$f_1(p) = e^{\frac{j}{2}\frac{d-1}{b}p^2 \Delta_t^2} f(p\Delta_t),$$

$$f_2(q) = \sum_{p=-M}^{M} e^{\frac{j}{b}(q\Delta_u - p\Delta_t)^2} f_1(p), \quad \text{and}$$

(124)

$$F_{(a,b,c,d)}(q\Delta_u) = \sqrt{\frac{1}{j2\pi b}} \cdot e^{\frac{j}{2}\frac{d-1}{b}q^2 \Delta_u^2} \cdot f_2(q).$$

The chirp convolution method has an important advantage: There is no constraint for the product of Δ_t and Δ_u.

5.2 DFT-like Implementation Method

Pei and Ding (2000a) proposed another method to implement the LCT. It is easy to prove that if $b \neq 0$, then the ABCD matrix of the LCT can be decomposed into

$$\begin{bmatrix} a & b \\ c & d \end{bmatrix} = \underbrace{\begin{bmatrix} 1 & 0 \\ d/b & 1 \end{bmatrix}}_{\text{(chirp multiplication)}} \cdot \underbrace{\begin{bmatrix} 0 & 1 \\ -1 & 0 \end{bmatrix}}_{\text{(FT)}} \cdot \underbrace{\begin{bmatrix} 1/b & 0 \\ 0 & b \end{bmatrix}}_{\text{(scaling)}} \underbrace{\begin{bmatrix} 1 & 0 \\ a/b & 1 \end{bmatrix}}_{\text{(chirp multiplication)}} \quad (b \neq 0).$$

(125)

Thus, if $b \neq 0$, we use the following steps to calculate the LCT with the parameters $\{a, b, c, d\}$:

    1. Chirp multiplication: $f_1(t) = \exp(jat^2/2b) \cdot f(t).$ (126)

    2. Scaling: $f_2(t) = \sqrt{b} \cdot f_1(b \cdot t) = \sqrt{b} \cdot e^{\frac{jab}{2}t^2} \cdot f(bt).$ (127)

    3. Fourier transform: $F_3(u) = \dfrac{1}{\sqrt{j2\pi}} \cdot \displaystyle\int_{-\infty}^{\infty} e^{-j \cdot u \cdot t} \cdot f_2(t) \cdot dt.$ (128)

    4. Chirp multiplication: $F_4(u) = \exp(jdu^2/2b) \cdot F_3(u).$ (129)

We can use this process to implement the LCT. In the third step, if we sample t and u as $p\Delta_t$ and $q\Delta_u$, and Δ_t and Δ_u satisfy the constraint that

$$\Delta_t \Delta_u = 2\pi/P (\text{with } P \text{ as an integer}), \qquad (130)$$

then it will become the DFT. Therefore, we can implement the LCT as the DFR-like method, as follows:

$$F_{(a,b,c,d)}(s\Delta_u) = \Delta_t \sqrt{\frac{b}{j2\pi}} e^{j\frac{d}{2b} \cdot q^2 \Delta_u^2} \sum_{p=-M}^{M} e^{-j\frac{2\pi pq}{P}} \left[e^{j\frac{ab}{2}p^2\Delta_t^2} f(p \cdot b \cdot \Delta_t) \right], \qquad (131)$$

where $M = (P-1)/2$ and $\Delta_t\Delta_u = 2\pi/P$.

When using the DFT-like method to implement the LCT, two P-point chirp multiplication operations and one P-point DFT are required. Since the P-point DFT requires about $(P/2)\log_2 P$ multiplications (Burrus, 1977), the complexity of implementation is

$$2P(\text{two multiplication operations})$$
$$+ (P/2)\log_2 P(\text{one DFT}) \approx (P/2) \cdot \log_2 P. \qquad (132)$$

The DFT-like implementation method has less computation loading, but there is a constraint for the product of Δ_t and Δ_u in Eq. (130).

Note that when P is large, the number of multiplications of the LCT is almost the same as the number of multiplications required for implementing the P-point DFT [i.e., $(P/2)\log_2 P$]. In other words, the computation loading of the LCT is very close to that of the DFT. However, the LCT is much more flexible and can achieve better performance in signal processing applications.

5.3 Discrete Versions of the LCT

It is well known that the discrete version of the FT is called the *discrete Fourier transform (DFT)*. The discrete version of the FRFT [i.e., the *discrete fractional Fourier transform (DFRFT)*] has also been derived (Santhanam & McClellan, 1996; Arikan et al., 1996; Pei & Yeh, 1997; Pei, Yeh, & Tseng, 1999; Pei, Hsue, & Ding, 2006; Candan, 2007; Pei & Ding, 2008). The DFRFT was derived based on decomposing the eigenfunctions and fractionalizing the eigenvalues of the DFT. The derived DFRFT has the additivity property, and its properties are close to those of the DFT.

However, since the LCT has four parameters, finding the discrete linear canonical transform (DLCT) that satisfies the additivity property in Eqs. (9) and (10) is very difficult. Pei and Ding (2000a) treat Eq. (131) as the discrete version of the LCT. However, it does not have the additivity property.

Ding and Pei (2011b) proposed a way to define the DLCT with the additivity property. They applied the discrete-time FT and bilinear mapping. The process of the DLCT is as follows:

Step 1. Perform the discrete-time FT (DTFT) for $x[n]$:

$$X(\omega) = \sum_n x[n]e^{-j\omega n}, \tag{133}$$

where $-\pi < \omega < \pi$.

Step 2. Convert $X(\omega)$ into $X_1(\omega)$:

$$X_1(\omega) = \sqrt{\phi\prime(\omega)}X(\phi(\omega)), \tag{134}$$

where $\phi(\omega)$ is a one-to-one mapping operation, $-\infty < \omega < \infty$, and $-\pi < \phi(\omega) < \pi$. In particular, if

$$\phi(\omega) = 2\operatorname{atan}\left(\frac{\omega}{2}\right), \tag{135}$$

where *atan* refers to the arctangent, then Eq. (134) becomes the bilinear transform:

$$X_1(\omega) = \sqrt{\frac{4}{\omega^2 + 4}}X(2\operatorname{atan}(\omega/2)). \tag{136}$$

Step 3. Perform the continuous LCT with parameters $\{d, -c, -b, a\}$ for $X_1(\omega)$:

$$X_2(\rho) = O_F^{(d,-c,-b,a)}[X_1(\omega)] = \sqrt{\frac{1}{j2\pi b}} \int_{-\infty}^{\infty} e^{-j\frac{d}{2c}\rho^2 + j\frac{1}{c}\rho\omega - j\frac{d}{2c}\omega^2} X_1(\omega)d\omega. \tag{137}$$

Step 4. Convert $X_2(\rho)$ into $Y(\rho)$, where

$$Y(\rho) = \sqrt{\frac{1}{\phi\prime(\phi^{-1}(\rho))}}X_2\left(\phi^{-1}(\rho)\right), \tag{138}$$

$-\pi < \rho < \pi$ and $-\infty < \phi^{-1}(\rho) < \infty$. In particular, if we choose $\phi(\omega)$, as in Eq. (135), then

$$Y(\omega) = \left|\sec\left(\frac{\omega}{2}\right)\right| X_2\left(2\tan\left(\frac{\omega}{2}\right)\right). \qquad (139)$$

Step 5. Then, perform the inverse discrete-time Fourier transform (IDTFT) for $Y(\rho)$:

$$y[n] = \frac{1}{2\pi}\int\limits_{-\pi}^{\pi} Y(\omega)e^{j\omega n}d\omega. \qquad (140)$$

The output $y[n]$ is the DLCT of $x[n]$, and is denoted as follows:

$$y[n] = O_{DLCT}^{(a,b,c,d)}(x[n]). \qquad (141)$$

Ding and Pei (2011b) proved that the DLCT defined in Eq. (141) has the following additivity property:

$$O_{DLCT}^{(a_3,b_3,c_3,d_3)}(x[n]) = O_{DLCT}^{(a_2,b_2,c_2,d_2)}\left\{O_{DLCT}^{(a_1,b_1,c_1,d_1)}(x[n])\right\}, \qquad (142)$$

where $\{a_1, b_1, c_1, d_1\}$, $\{a_2, b_2, c_2, d_2\}$, and $\{a_3, b_3, c_3, d_3\}$ have the same relation as in Eq. (10). The DLCT is useful for designing the digital filter and defining the discrete scaling operation with the additivity property.

6. APPLICATIONS OF THE LINEAR CANONICAL TRANSFORM IN SIGNAL PROCESSING

Figure 2.2 summarizes the applications of the LCT. In addition to computing the convolution operation, almost all the applications of the FT are the applications of LCT. Furthermore, since the LCT is more general and flexible than the FT, using the LCT instead of the LCT in these applications can achieve even better performance.

Due to the space limitations, in the rest of this section, we discuss the applications of the LCT for filter design (including the deterministic case and the random process case), signal sampling, modulation and multiplexing, bandwidth saving, and image processing (including space-variant pattern recognition, asymmetric edge detection, and encryption). The applications of the LCT for electromagnetic wave propagation analysis will be introduced in the section "Linear Canonical Transform for Electromagnetic Wave Propagation Analysis," later in this chapter.

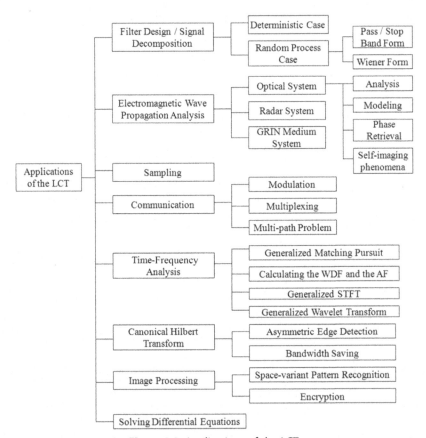

Figure 2.2 Applications of the LCT.

6.1 Application of the LCT for Filter Design in the Deterministic Case

It is well known that the conventional filter can be expressed as a convolution operation and can be implemented by the FT. Mendlovic *et al.* (1996), Erden and Ozaktas (1998), and Pei and Ding (2007) have applied the FRFT to filter design and showed that the FRFT is very helpful for removing noise whose instantaneous frequency varies with time. In fact, the LCT is also very useful for filter design, which was discussed in Pei and Ding (2001a).

The conventional filter can be expressed as

$$x_o(t) = \int\limits_{-\infty}^{\infty} h(t-\tau)x_i(\tau)d\tau, \qquad (143)$$

where $x_i(t)$ is the input, $x_o(t)$ is the output, and $h(t)$ is the impulse response of the system. It can also be written as

$$x_o(t) = \frac{1}{\sqrt{2\pi}} IFT(FT(x_i(t)) \cdot H(\omega)), \quad \text{where } H(\omega) = FT(h(t)). \quad (144)$$

The LCT can be used to generalize the conventional filter. In Eq. (144), the FT can be replaced by the LCT with parameters $\{a, b, c, d\}$, and the IFT can be replaced by the LCT with parameters $\{d, -b, -c, a\}$. Then the canonical filter is defined as

$$x_o(t) = O_F^{(d,-b,-c,a)} \left(O_F^{(a,b,c,d)}(x_i(t)) \cdot H(u) \right), \quad (145)$$

where $H(u)$ is the transfer function. $H(u)$ can be designed as a pass–stop band:

$$H(u) = 1 \text{ for the pass band; } H(u) = 0 \text{ for the stop band.} \quad (146)$$

Note that the canonical filter is a special case of the canonical convolution introduced in the section "Basic Properties and Transform Pairs of Linear Canonical Transforms," earlier in this chapter.

The canonical filter can remove some noises that are very hard to remove by a conventional filter, such as chirp noise:

$$n(t) = C \cdot \exp\left(j\left(h_1 t^2 + h_2 t + h_3\right)\right). \quad (147)$$

This type of noise often occurs in the optical system, the microwave system, the radar system, and acoustics. Note that the WDF of $n(t)$ in (147) is

$$W_n(t,f) = 2\pi|C|^2 \cdot \delta(f - 2h_1 t - h_2). \quad (148)$$

After performing the LCT, from Eq. (33), the WDF of the LCT of $n(t)$ is

$$W_{N_{(a,b,c,d)}}(u, v) = 2\pi|C|^2 \cdot \delta(-cu + av - 2h_1(du - bv) - h_2). \quad (149)$$

One can choose a and b properly, such that

$$a = -2h_1 b. \quad (150)$$

Then,

$$W_{N_{(a,b,c,d)}}(u, v) = 2\pi|C|^2 \cdot \delta\left((-c - 2h_1 d)\left(u + \frac{h_2}{c + 2h_1 d}\right)\right). \quad (151)$$

Therefore, one can use the canonical filter in Eq. (145), where a and b satisfy Eq. (150) and

$$H(u) = 0 \text{ when } u \text{ is near to } \frac{-h_2}{c + 2h_1 d}, \text{ and } H(u) = 1 \text{ otherwise}, \quad (152)$$

to remove the noise in Eq. (147). In addition to chirp noise and a linear combination of chirps, the canonical filter with the pass-stop band transfer function in Eq. (146) are also very helpful for removing the noise whose time-frequency distribution does not overlap with that of the signal.

The time-frequency distributions, including the WDF, the short-time FT (Bastiaans, 1980), and the Gabor-Wigner transform (Pei & Ding, 2007), are helpful for designing the canonical filter (Pei & Ding, 2001a). In general,

1. The parameter $-a/b$ of LCT is the slope of the cutoff line on the time-frequency plane.
2. The transfer function $H(u)$ can control the location of the pass region on the time-frequency plane.

For example, suppose that the time-frequency distribution of a received signal is as shown in Figure 2.3. The dotted regions mean the locations where

$$W_x(t, \omega) > T, \quad (153)$$

where T is the threshold.

We also suppose that the desired and undesired parts can be separated by two parallel cutoff lines. Then we can choose $\{a, b, c, d\}$ and $H(u)$ of the canonical filter as

$$a/b = \omega_1/t_1, \quad H(u) = \Pi\left(\frac{u - a(t_0 + t_1)/2}{a(t_0 - t_1)}\right). \quad (154)$$

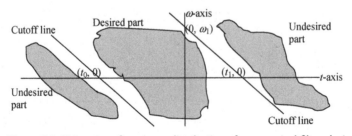

Figure 2.3 Using time-frequency distributions for canonical filter design.

If we must use several cutoff lines with different slopes to separate the desired and undesired parts on the time-frequency plane, then we must also repeat the pass-stop band canonical filters several times to remove the noise. For each time, we choose the values of a/b and $H(u)$ properly to control the slope and the location of the cutoff line on the time-frequency plane.

Since the signal part and the noise part are easier to separate in the time-frequency plane (two dimensions) than in the ω-axis (one dimension), using the canonical filter can remove more noise than using a conventional filter.

6.2 Application of the LCT for Filter Design in the Random Process Case

When the noise does not have a deterministic form, the Wiener filter can be used to remove the noise. Zalevsky and Mendlovic (1996) and Kutay *et al.* (1997), they generalized the Wiener filter to the case of the FRFT. Moreover, Barshan, Kutay, and Ozaktas (1997) and Scharf and Thomas (1998) further generalized the Wiener filter to the case of the LCT and defined the Wiener filter in canonical coordinates. We call this the *canonical Wiener filter.*

The method to design the canonical Wiener filter is described as follows. Suppose a cross-correlation between the original signal $s(t)$ and the received signal $x_i(t)$, denoted by $R_{si}(t, \sigma)$); an auto-correlation of the received signal (denoted by $R_{ii}(t, \sigma)$); and an auto-correlation of the signal (denoted by $R_{ss}(t, \sigma)$).

Then the transfer function $H(u)$ of the canonical filter in Eq. (145) can be designed as

$$H_{opt}(u) = R_{S,I}(u, u) / R_{I,I}(u, u), \tag{155}$$

where $R_{S,I}(u, u)$ and $R_{I,I}(u, u)$ are calculated as follows:

$$R_{S,I}(u, u) = \int\limits_{-\infty}^{\infty} \int\limits_{-\infty}^{\infty} K_{(a,b,c,d)}(u, t) K_{(a,b,c,d)}^{*}(u, \sigma) R_{si}(t, \sigma) dt \, d\sigma, \tag{156}$$

$$R_{I,I}(u, u) = \int\limits_{-\infty}^{\infty} \int\limits_{-\infty}^{\infty} K_{(a,b,c,d)}(u, t) K_{(a,b,c,d)}^{*}(u, \sigma) R_{ii}(t, \sigma) dt \, d\sigma, \tag{157}$$

$$K_{(a,b,c,d)}(u, t) = \sqrt{\frac{1}{j2\pi b}} \cdot e^{j\frac{d}{2b}u^2} \cdot e^{-j\frac{1}{b}ut} \cdot e^{j\frac{a}{2b}t^2}. \tag{158}$$

This is the method of designing the canonical Wiener filter. Moreover, the mean square error (MSE) of the reconstructed signal can be calculated as follows:

$$MSE = \int\limits_{-\infty}^{\infty} \left[R_{S,S}(u,u) - 2\mathrm{Re}\left(H_{opt}^*(u) R_{S,I}(u,u) \right) \right.$$

$$\left. + \left| H_{opt}(u) \right|^2 R_{I,I}(u,u) \right] \cdot du, \tag{159}$$

where

$$R_{S,S}(u,u) = \int\limits_{-\infty}^{\infty} \int\limits_{-\infty}^{\infty} K_{(a,b,c,d)}(u,t) K_{(a,b,c,d)}^*(u,\sigma) R_{ss}(t,\sigma) dt \, d\sigma. \tag{160}$$

When $\{a, b, c, d\}$ is given, Eq. (155) can be used to determine the optimal filter and use Eq. (159) to calculate the MSE of the reconstructed signal. Moreover, the parameter set $\{a, b, c, d\}$ can be varied iteratively, and Eq. (159) can be used to determine the optimal parameter set $\{a, b, c, d\}$ such that the MSE can be minimized.

Furthermore, Pei and Ding (2010) proposed a way to minimize the effect of white noise. If $n(t)$ is white noise, then the mean of the WDF of $n(t)$ is a constant (Martin & Flandrin, 1983):

$$E[W_n(t,f)] = \sigma, \tag{161}$$

where σ is some constant. Therefore, to reduce the effect of the white noise, one must perform the canonical filter iteratively to generate the cutoff lines in the time-frequency plane [the relations between the slope and the location of the cutoff line and the parameters of the canonical filter have been discussed in Figure 2.3 and Eq. (154)]. The pass region should be as close to the signal part in the time-frequency plane as possible, as shown in Figure 2.4. If the area of the pass region is A, the signal-to-noise ratio (SNR) of the reconstructed signal is

$$SNR \approx \log_{10} \frac{E_x}{\sigma A}, \tag{162}$$

where $E_x = \int_{-\infty}^{\infty} |x(t)|^2$ and $x(t)$ is the signal. According to the conventional communication theory, to reduce the effect of the white noise, the *bandwidth* of a signal should be as narrow as possible. Now, with the canonical filters, from Figure 2.4 and Eq. (162), the theorem can be

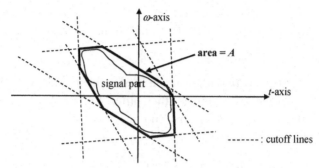

Figure 2.4 Reducing the effect of the white noise iteratively by canonical filters.

modified. To reduce the effect of the white noise, *the area of the time-frequency distribution* of a signal should be as small as possible.

6.3 Generalized Sampling Theorem Using the LCT

Xia (1996) generalized the conventional sampling theory to apply it to the case of the FRFT. He found that if the input signal is band-limited in the FRFT domain, then if the sampling interval is chosen properly, then the original signal can still be retrieved. Ozaktas and Arikan (1996), Marinho and Bernardo (1998), and Ding and Pei (2005) discussed the best way to choose the sampling interval for the signal that is limited in the FRFT properly. Then, Stern (2006), Tao *et al.* (2008), and Healy and Sheridan (2009) further generalized the sampling theory by the LCT. In this section, we describe the sampling theory in the LCT domain.

Assume that $f(t)$ is band-limited (or almost band-limited) in the transformed domain corresponding to the LCT with parameters $\{a, b, c, d\}$:

$$F_{(a,b,c,d)}(u) = 0 \quad \text{for } |u| > \Omega. \tag{163}$$

Then $f(t)$ can be recovered from $F_{(a,b,c,d)}(u)$ by the LCT with parameters $\{d, -b, -c, a\}$:

$$f(t) = \sqrt{\frac{j}{2\pi b}} \cdot e^{-\frac{j}{2}\frac{a}{b}t^2} \int_{-\Omega}^{\Omega} e^{\frac{j}{b}tu} e^{-\frac{j}{2}\frac{d}{b}u^2} F_{(a,b,c,d)}(u) \cdot du. \tag{164}$$

The integration part of Eq. (164) is

$$g(t) = \sqrt{-jb} \cdot e^{\frac{j}{2}\frac{a}{b}t^2} f(t) \cong \frac{1}{\sqrt{2\pi}} \int_{-\Omega}^{\Omega} e^{\frac{j}{b}tu} e^{-\frac{j}{2}\frac{d}{b}u^2} F_{(a,b,c,d)}(u) \cdot du. \tag{165}$$

It is easy to prove that $g(t)$ is band-limited in the conventional frequency domain. If $G(\omega)$ is the FT of $g(t)$, then one can prove that

$$G(\omega) = 0 \quad \text{for } |\omega| > \Omega/|b|. \tag{166}$$

Therefore, although $f(t)$ is not band-limited in the frequency domain, if it is band-limited in the LCT domain, then $g(t)$ in Eq. (165) is band-limited. We can use this fact to develop the sampling theorem of the LCT. If we know the value of

$$g(n\Delta) = \sqrt{-j2\pi b} \cdot e^{\frac{j}{2}\frac{d}{b}n^2\Delta^2} f(n\Delta), \tag{167}$$

where $n \in Z$ and $\Delta < \pi|b|/\Omega$, and

$$G_{sp}(\omega) = \frac{1}{\sqrt{2\pi}} \sum_n e^{-j\omega n\Delta_t} g(n\Delta_t) \cdot \Delta_t, \tag{168}$$

then

$$G(\omega) = G_{sp}(\omega) \quad \text{for } |\omega| \le \Omega/|b| \tag{169}$$

is satisfied. Therefore, one can use the following equation to obtain $F_{(a,b,c,d)}(u)$:

$$F_{(a,b,c,d)}(u) = e^{\frac{j}{2}\frac{d}{b}u^2} G(u/b). \tag{170}$$

To reconstruct the original signal, one can perform the low-pass filter on $g(n\Delta_t)$ (i.e., convolving $g(n\Delta_t)$ with a sinc function) and use the relation in Eq. (167) to retrieve $f(t)$.

In summary, if $f(t)$ is band-limited in the LCT domain and its LCT satisfies Eq. (163), then we can sample $f(t)$ by the interval satisfying the following equation:

$$\Delta < \pi|b|/\Omega, \tag{171}$$

and we can recover $f(t)$ from $\{f(n \cdot \Delta) \mid n \in Z\}$ by a conventional low-pass filter:

$$f(t) = e^{-j\frac{d}{2b}t^2} \cdot IFT\left(FT\left(e^{j\frac{d}{2b}t^2} \cdot f_s(t) \right) \cdot \Pi\left(\frac{\omega}{2\omega_s} \right) \right), \tag{172}$$

where

$$f_s(t) = f(n\Delta) \cdot \delta(t - n\Delta) \quad \text{when } t = n \cdot \Delta, \text{ with } n \text{ being any integer; and}$$
$$f_s(t) = 0; \qquad\qquad\qquad \text{otherwise;}$$

$$\tag{173}$$

$$\Pi\left(\frac{\omega}{2\omega_s}\right) = 1 \text{ if } |\omega| \leq \omega_s, \Pi\left(\frac{\omega}{2\omega_s}\right) = 0;$$

$$\text{otherwise,} \quad \frac{\Omega}{b} < \omega_s < \frac{2\pi}{\Delta} - \frac{\Omega}{b}. \tag{174}$$

Eq. (172) can also be rewritten for the generalized Shannon theorem as follows:

$$f(t) = e^{\frac{-j}{2}\frac{a}{b}\cdot t^2} \cdot \sum_{k=-\infty}^{\infty} \left[\frac{\Delta}{\pi(t - k\Delta)} \cdot f(k\Delta) \cdot e^{\frac{j}{2}\frac{a}{b}\cdot (k\Delta)^2} \cdot \sin(\omega_s \cdot (t - k\Delta))\right], \tag{175}$$

With the aid of generalized sampling/Shannon theorems, many signals that are not band-limited in the frequency domain become recoverable after sampling by the LCT. Generalized sampling/Shannon theorems are useful for communication.

6.4 Modulation by the Linear Canonical Transform

It is common to use the conventional FT for modulation. Ozaktas *et al.* (1994) and Mendlovic and Lohmann (1997) used the FRFT for signal modulation to improve the channel capacity. In fact, the LCT is also useful for modulation, as we discuss next.

When using the conventional method for modulation, if there are N signals $\{g_1(t), g_2(t), \ldots\ldots, g_N(t)\}$ and these signals are band-limited in the frequency domain,

$$G_n(\omega) \approx 0 \quad \text{for} |w| > B_n, n = 1, 2, \ldots, N, G_n(\omega) = FT(g_n(t)), \tag{176}$$

then these signals can be modulated as

$$g_M(t) = e^{j\omega_0 t} g_0(t) + e^{j\omega_1 t} g_1(t) + e^{j\omega_2 t} g_2(t) + \ldots\ldots + e^{j\omega_N t} g_N(t), \tag{177}$$

where

$$\omega_n - B_n > \omega_{n-1} + B_{n-1} \text{ for } n = 1, 2, \ldots, N. \tag{180}$$

However, when the supports of some signals are not limited in the frequency domain but are limited in the LCT domain, it is proper to use the LCT for modulation.

When using the LCT for modulation, we can apply the following process:

Step 1: For all the input signals $\{g_1(t), g_2(t), \ldots\ldots\ldots, g_N(t)\}$, find the parameters $\{a_n, b_n, c_n, d_n\}$, such that the LCT with parameters $\{a_n, b_n, c_n, d_n\}$ for $g_n(t)$ has limited support:

$$G_{n(a_n,b_n,c_n,d_n)}(u) \approx 0 \quad \text{for } |u| > D_n, \tag{181}$$

where

$$G_{n(a_n,b_n,c_n,d_n)}(u) = O_F^{(a_n,b_n,c_n,d_n)}(g_n(t)),$$

and the value of D_n should be as small as possible.

Step 2: Then, perform the LCT with parameters $\{-c_n, -d_n, a_n, b_n\}$ for $g_n(t)$, and obtain $f_n(t)$:

$$f_n(t) = O_F^{(-c_n,-d_n,a_n,b_n)}(g_n(t)). \tag{182}$$

Note that the LCT with parameters $\{-c_n, -d_n, a_n, b_n\}$ is just the IFT of the LCT with parameters $\{a_n, b_n, c_n, d_n\}$. Therefore, $f_n(t)$ has limited support in the frequency domain.

Step 3: Then we can use the conventional modulation algorithm to modulate $\{f_1(t), f_2(t), \ldots\ldots\ldots, f_N(t)\}$:

$$f_M(t) = e^{j\omega_0 t}f_0(t) + e^{j\omega_1 t}f_1(t) + e^{j\omega_2 t}f_2(t) + \ldots\ldots + e^{j\omega_N t}f_N(t), \tag{183}$$

where

$$\omega_n - D_n > \omega_{n-1} + D_{n-1} \quad \text{for } n = 1, 2, \ldots., N, \tag{184}$$

and the total bandwidth is

$$W > 2(D_0 + D_1 + D_2 + \ldots\ldots + D_N). \tag{185}$$

When performing demodulation, one can just apply the process as follows:

Step 1: Perform demultiplexing for $f_M(t)$:

$$f_n(t) = LP_{\omega_{n,o}}\left(e^{-j\omega_n t}f_M(t)\right), \tag{186}$$

where

$$\omega_{n+1} - \omega_n - D_{n+1} > \omega_{n,o} > D_n,$$

$$LP_{\omega_{n,o}}(h(t)) = IFT\left(FT(h(t)) \cdot \Pi(\omega/2\omega_{n,o})\right),$$

$$\Pi(\omega/2\omega_{n,o}) = 1 \quad \text{for} |\omega| < \omega_{n,0}, \text{and } \Pi(\omega/2\omega_{n,o}) = 0 \text{ otherwise.}$$

$$\tag{187}$$

Step 2: Then, $g_n(t)$ can be recovered from

$$g_n(t) = O_F^{(b_n,d_n,-a_n,-c_n)}\left(f_n\{t\}\right). \tag{188}$$

Using the LCT for modulation is especially useful for signals whose instantaneous frequency varies with time, such as the acoustic signal and the radar signal. With the aid of the LCT, the capacity of a channel can be improved.

Furthermore, when $g_n(t)$ is real, one can use the canonical Hilbert transform (as described in the section "Canonical Hilbert Transform," earlier in this chapter) to generate the analysis signal and save half of the bandwidth (Pei & Ding, 2003c). It can further improve the efficiency of signal transmission.

6.5 Linear Canonical Transform for Image Processing

There are at least three possible applications of the LCT for image processing, including space variant pattern recognition, asymmetric pattern recognition, and encryption. It is well known that one can apply the FT for calculating the correlation and performing space-invariant pattern recognition. Lohmann, Zalevsky, and Mendlovic (1996), Granieri, Arizaga, and Sicre (1997), and Kutay and Ozaktas (1998) generalized correlation by the FRFT and used it for space-variant pattern recognition. Pei and Ding (2000a) further generalize these works and applied the canonical correlation (see the section "Canonical Convolution and Canonical Correlation," earlier in this chapter) for space-variant pattern recognition. When applying space-variant pattern recognition, an object can be detected only when it both matches the reference pattern and is located in a desired region.

The canonical Hilbert transform described in the section "Canonical Hilbert Transform" is useful for asymmetric edge detection. It is a generalization of the Hilbert transform and the fractional Hilbert transform (Lohmann, Mendlovic, & Zalevsky, 1996; Zayed, 1998). For the application of asymmetric edge detection, the canonical Hilbert transform in Eq. (86) is generalized as

$$x_H^{(a,b,c,d,\sigma,\tau)}(t) = O_F^{(d,-b,-c,a)}\left[O_F^{(a,b,c,d)}(x(t))\{\sigma + \tau H(u)\}\right], \tag{189}$$

where $H(u)$ is defined in Eq. (85). The conventional Hilbert transform treat the rising edge and the falling edge equally [the rising edge means that $x'(t) > 0$, and the falling edge means that $x'(t) < 0$]. When using the canonical

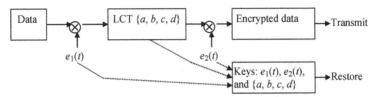

Figure 2.5 Process of encryption by the LCT.

Hilbert transform, if $\sigma < 0$, the rising edge will be further emphasized. If $\sigma >$ 0, the falling edge will be further emphasized. Furthermore, in the two-dimensional case, one can apply the canonical Hilbert transform to perform directional edge detection (Pei & Ding, 2003b); i.e., detecting the edge with a certain direction.

The LCT can also be applied for image encryption. Using the LCT for encryption is more effective than using the LCT since the LCT has four parameters, which can be treated as the extra key to improve the safety. If the parameters of the LCT are not known, one cannot retrieve the original image. Zalevsky, Mendlovic, and Dorsch (1996), Cong, Chen, and Gu (1998), and Liu, Yu, and Zhu (2001) applied the FRFT for image encryption. Hennelly and Sheridan (2004) applied the Fresnel transform for encryption. Singh and Sinha (2010) applied the LCT for image encryption.

One of the possible structures that use the LCT for image encryption is plotted in Figure 2.5. If one wants to recover the encrypted data, then one must know $e_1(t)$, $e_2(t)$, and the parameters $\{a, b, c, d\}$ of the LCT. Otherwise, the original image cannot be retrieved.

The encryption structure in Figure 2.5 is basic. We can replace the two multiplications in Figure 2.5 with other reversible operations, repeat this structure several times, or change this structure into multiple channels, among other options.

7. LINEAR CANONICAL TRANSFORM FOR ELECTROMAGNETIC WAVE PROPAGATION ANALYSIS

Nazarathy and Shamir (1982), Bastiaans (1991), Abe and Sheridan (1994b), Ozaktas and Mendlovic (1996), and Bernardo (1996) have found that the LCT and the offset LCT defined in Eqs. (21) and (22) are useful for optical system analysis because many monochromic light propagation operations in optical systems can be described by LCT. In addition, the LCT is useful for radar system analysis, gradient-index fiber system analysis, and analyzing other electromagnetic wave propagation systems. We discuss how to apply the LCT for these applications in this section.

7.1 Using the LCT to Represent Optical Components

Many optical components can be modeled by the LCT or the offset LCT. We describe some of them in this subsection.

a Propagation Through the Cylinder Lens with Focal Length f

Suppose that a monochromatic light with wavelength λ enters a cylinder lens and its field distribution is $U_i(x)$. Also assume that the cylinder lens has a focal length of f, thickness of Δ, and refractive index of η. To make a distinction, we use η instead of n to denote the refractive index and use n to denote the sixth parameter of the offset LCT (see Eq. (20). Then, from paraxial approximation, the output will have the distribution $U_o(x)$ (Goodman, 2005):

$$U_o(x) = e^{j2\pi\eta\Delta/\lambda} \cdot e^{-j\frac{\pi}{\lambda f} \cdot x^2} \cdot U_i(x). \tag{190}$$

If the difference of the constant phase is ignored, we find that Eq. (190) is just the LCT with the following parameters:

$$\begin{bmatrix} a & b \\ c & d \end{bmatrix} = \begin{bmatrix} 1 & 0 \\ -2\pi/\lambda f & 1 \end{bmatrix}. \tag{191}$$

b Propagation Through the Free Space (Fresnel Transform) with Length z

Suppose that a monochromatic light has the wavelength of λ and the distribution of $U_i(x)$. If it propagates through a free space with distance z, then, from the Fresnel theory (Goodman, 2005), the output $U_o(s)$ is

$$U_o(s) = \frac{e^{j2\pi \cdot z/\lambda}}{j\lambda z} \cdot \int\limits_{-\infty}^{\infty} e^{j\frac{\pi}{\lambda z}(s-x)^2} \cdot U_i(x) \cdot dx. \tag{192}$$

Eq. (192) is called the *Fresnel transform*. Comparing it with Eq. (6), if the difference of the constant phase is ignored, it just corresponds to the LCT with the following parameters:

$$\begin{bmatrix} a & b \\ c & d \end{bmatrix} = \begin{bmatrix} 1 & \lambda z/2\pi \\ 0 & 1 \end{bmatrix}. \tag{193}$$

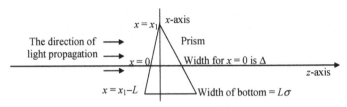

Figure 2.6 A prism that has a height of L and a bottom of width $L\sigma$.

The light propagation through a prism, as in Figure 2.6, can be derived as follows:

$$O^{\theta}_{prism}(g(x)) = e^{j\frac{2\pi}{\lambda}\eta\Delta}e^{-j\frac{2\pi}{\lambda}(\eta-1)\sigma x}g(x), \tag{194}$$

where $\sigma = $ (bottom width)/(height), η is the refractive index, and Δ is the thickness at $x = 0$. It is a special case of the offset LCT [defined in Eqs. (20) and (21)] with the following parameters:

$$\begin{bmatrix} a & b \\ c & d \end{bmatrix} = \begin{bmatrix} 1 & 0 \\ 0 & 1 \end{bmatrix}, \quad \begin{bmatrix} m \\ n \end{bmatrix} = \begin{bmatrix} 0 \\ -2\pi(\eta-1)\sigma/\lambda \end{bmatrix}. \tag{195}$$

The light propagation through a cylinder lens with focal length f, but with the lens shifted upward at a distance of x_0, as in Figure 2.7, is derived as follows:

$$O^{f,x_0}_{lens}(g(x)) = e^{j\frac{2\pi}{\lambda}\eta\Delta}e^{-j\frac{\pi}{f\cdot\lambda}(x-x_0)^2}g(x) = e^{j\frac{\pi}{f\lambda}\left(2f\eta\Delta-x_0^2\right)}e^{-j\frac{\pi}{f\cdot\lambda}x^2}e^{j\frac{2\pi}{f\cdot\lambda}x_0\cdot x}g(x), \tag{196}$$

where η is the refractive index. It corresponds to the offset LCT with the following parameters:

$$\begin{bmatrix} a & b \\ c & d \end{bmatrix} = \begin{bmatrix} 1 & 0 \\ -2\pi/f\lambda & 1 \end{bmatrix}, \quad \begin{bmatrix} m \\ n \end{bmatrix} = \begin{bmatrix} 0 \\ 2\pi x_0/f\lambda \end{bmatrix}. \tag{197}$$

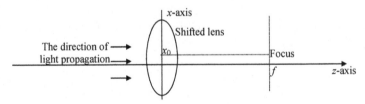

Figure 2.7 A lens with an upward shift of x_0.

There are many other optical propagation operations that can be represented by the LCT. Note that among the abovementioned optical propagation operations, the free space propagation operation and the light propagation operation through a cylinder lens without shifting can be represented by the LCT with only four parameters $\{a, b, c, d\}$, but the other two operations need to be modeled by the offset LCT with six parameters $\{a, b, c, d, m, n\}$.

7.2 Using the LCT or Offset LCT to Represent the Optical Systems

Since the LCT can model all four optical operations described in the previous section, if an optical system is composed of these components, we can use the LCT to represent the optical system. The process is as follows:

1. For each component in the optical system, find the parameters of the LCT or offset LCT that can represent it. Then we can use an ABCD-MN matrix to represent each of the components.
2. Then, after calculating the product of ABCD-MN matrices, one can obtain the parameters of the LCT or offset LCT that can represent the whole optical system.

Since we can use the offset LCT to represent many optical systems, the offset LCT can be applied to solve or illustrate many problems associated with the optics. We often analyze and illustrate the optical system problems by calculating the integral equations or other complex equations. Now, with the aid of the offset LCT and its ABCD-MN matrix representation, many complicated problems in optics can be solved easily by just performing the matrix operation.

To illustrate this, note the following example. We will use the offset LCT to illustrate the deflection phenomenon caused by the prism. We know that the direction of the light propagation will change if the light passes through a prism, as shown in Figure 2.8. Although integral equations also can be used to prove this phenomenon, the proof is very complicated. Here, we use the ABCD-MN matrix of the offset LCT to explain the deflection phenomenon.

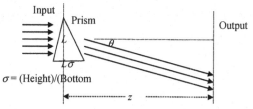

Figure 2.8 The deflection phenomenon of the prism can also be interpreted by the offset LCT.

The optical system in Figure 2.8 consists of two optical components: a prism and a free space. They can be represented by the LCT with the parameters as in Eqs. (195) and (193), respectively. Therefore, we can use the ABCD-MN matrix representation as follows to represent the system in Figure 2.8:

$$
\begin{bmatrix} 1 & \dfrac{\lambda z}{2\pi} \\ 0 & 1 \end{bmatrix} \cdot \left(\begin{bmatrix} 1 & 0 \\ 0 & 1 \end{bmatrix} \begin{bmatrix} t \\ k \end{bmatrix} + \begin{bmatrix} 0 \\ -\dfrac{2\pi(\eta-1)\sigma}{\lambda} \end{bmatrix} \right)
$$

$$
= \begin{bmatrix} 1 & \dfrac{\lambda z}{2\pi} \\ 0 & 1 \end{bmatrix} \begin{bmatrix} t \\ k \end{bmatrix} + \begin{bmatrix} -(\eta-1)\sigma z \\ \dfrac{2\pi(\eta-1)\sigma}{\lambda} \end{bmatrix}. \tag{198}
$$

Thus, if $g_1(x)$ is the input of the optical system in Figure 2.8 and $g_2(s)$ is the output, then

$$
g_2(s) = O_F^{(1,\lambda z/2\pi,0,1,-(\eta-1)\sigma z,-2\pi(\eta-1)\sigma/\lambda)}(g_1(x)). \tag{199}
$$

By contrast, if there is no prism, the output $g_3(s)$ is

$$
g_3(s) = O_F^{(1,\lambda z/2\pi,0,1,0,0)}(g_1(x)); \tag{200}
$$

i.e., $g_3(s)$ is the Fresnel transform of $g_1(x)$. Then the relation between $g_2(s)$ and $g_3(s)$ is

$$
\begin{aligned}
g_2(s) &= O_F^{(1,0,0,1,-(\eta-1)\sigma z,-\frac{2\pi}{\lambda}(\eta-1)\sigma)} \\
(g_3(s)) &= e^{-j\frac{2\pi}{\lambda}(\eta-1)\sigma s}g_3(s+(\eta-1)\sigma z).
\end{aligned} \tag{201}
$$

Comparing their intensities (in an optical system, only the intensity of the light distribution can be observed), we get

$$
|g_2(s)|^2 = |g_3(s+(\eta-1)\sigma z)|^2. \tag{202}
$$

That is, when one places a prism before a free space with length z, then the intensity of the output light distribution will shift with the amount of $-(\eta-1)\sigma\cdot z$. Note that the amount of shift is linearly proportional to z. Therefore, one can also state that the prism will deflect the direction of the light propagation with the angle of θ, where

$$
\theta = -\tan^{-1}((\eta-1)\sigma). \tag{203}
$$

Thus, we have used LCT to illustrate the deflection phenomenon of the prism in the optical system successfully by just using the offset LCT and the matrix operation. Therefore, the offset LCT is very useful for analyzing the optical systems.

Although one can also use Huygens's theorem and the Fresnel integral to analyze optical systems, they are hard to compute both manually and using the computer. Using the ABCD-MN matrix of the LCT can greatly simplify the work of optical system analysis.

Furthermore, Pei and Ding (2002a, 2003a) also have applied the ABCD-MN matrix of the offset LCT with the eigenfunctions of the offset LCT to examine the self-imaging phenomena of optical systems.

Using the LCT or offset LCT to represent the optical system is a paraxial approximation. That is, the result is accurate when $|u|$ (with u as the independent variable of the output function) is small (i.e., near the z-axis). The result may not be accurate when $|u|$ is large (i.e., far from the z-axis). Wolf and Krotzsch (1999) discuss the metaxial correction of the FRFT for optical system representation. Their result can also be applied when the LCT or offset LCT is used to represent an optical system.

7.3 Implementing the FRFT or LCT by Optical Systems

In addition to using the LCT for optical system analysis, conversely, one can use an optical device to implement the LCT. Since all of the LCT can be decomposed as the combination of chirp multiplications and chirp convolutions, one can use the optical system composed of lenses and free spaces to perform the LCT. Because

$$\begin{bmatrix} a & b \\ c & d \end{bmatrix} = \begin{bmatrix} 1 & 0 \\ (d-1)/b & 1 \end{bmatrix} \cdot \begin{bmatrix} 1 & b \\ 0 & 1 \end{bmatrix} \cdot \begin{bmatrix} 1 & 0 \\ (a-1)/b & 1 \end{bmatrix} \text{ if } b \neq 0, \quad (204)$$

$$\begin{bmatrix} a & b \\ c & d \end{bmatrix} = \begin{bmatrix} 1 & (a-1)/c \\ 0 & 1 \end{bmatrix} \cdot \begin{bmatrix} 1 & 0 \\ c & 1 \end{bmatrix} \cdot \begin{bmatrix} 1 & (d-1)/c \\ 0 & 1 \end{bmatrix} \text{ if } c \neq 0; \quad (205)$$

therefore, for the case where $b \neq 0$, one can implement the LCT by the method shown in Figure 2.9. When $c \neq 0$, one can implement the LCT by

Figure 2.9 Optical implementation of the LCT ($b \neq 0$) with two cylinder lenses and one free space.

the method shown in Figure 2.10. In Figures 2.9 and 2.10, the values of f_1, d_0, f_2, d_1, f_0, and d_2 are

$$\text{For Figure 2.9: } f_1 = \frac{2\pi b}{\lambda(1-a)}, \quad d_0 = \frac{2\pi b}{\lambda}, \quad f_2 = \frac{2\pi b}{\lambda(1-d)}. \quad (206)$$

$$\text{For Figure 2.10: } d_1 = \frac{2\pi(d-1)}{\lambda c}, \quad f_0 = -\frac{2\pi}{\lambda c}, \quad d_2 = \frac{2\pi(a-1)}{\lambda c}. \quad (207)$$

With these distances and focal lengths, the input and output in Figure 2.9 or Figure 2.10 have the following relation:

$$g_o(x) = \exp\left(-j\frac{2\pi L}{\lambda}\right)\left(O_F^{(a,b,c,d)}(g_i(x))\right), \quad (208)$$

where L is the total length of the system. (In Figure 2.9, $L \cong d_0$. In Figure 2.10, $L \cong d_1 + d_2$.) If we want the optical implementation of LCT to be shorter in length, then the method in Figure 2.9 is preferred. If we want to save the number of lenses, save the hardware cost, or avoid placing the lenses contacting to the input and output, then the method in Figure 2.8 is preferred.

In some situations, the values of d_0, d_1, or d_2 in Eqs. (206) and (207) will be negative. For example, in Figure 2.9, if

$$b < 0, \quad (209)$$

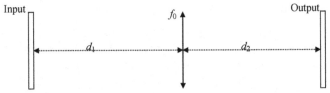

Figure 2.10 Optical implementation of the LCT ($c \neq 0$) with one cylinder lens and two free spaces.

then the value of d_0 is negative. In Figure 2.10, if one of the following conditions is satisfied:

$$(1)\ a > 1\ \text{or}\ d > 1\ \text{and}\ c < 0, \text{or}\ (2)\ a < 1\ \text{or}\ d < 1\ \text{and}\ c > 0, \quad (210)$$

then d_1, d_2, or both and either d_1 or d_2 will be negative. In these conditions, it is better to use another method to implement the LCT, or else implement the LCT with parameters $\{-a, -b, -c, -d\}$ instead of with parameters $\{a, b, c, d\}$. When implementing the LCT with parameters $\{-a, -b, -c, -d\}$, from Eq. (6), the intensity of the output is

$$\left| F_{(-a,-b,-c,-d)}(u) \right|^2 = \left| F_{(a,b,c,d)}(-u) \right|^2. \quad (211)$$

Remember that in an optical system, only the intensity can be observed. If we reverse the output, a valid result can be obtained.

7.4 Gradient–Index Medium System Analysis by the LCT

Mendlovic and Ozaktas (1993), Alieva *et al.* (1994), Mendlovic *et al.* (1994b), Yu *et al.* (1998), and Ozaktas *et al.* (1999) have found that if a gradient-index (GRIN) medium has the following refractive index distribution:

$$n^2(x, y) = n_0^2 \lfloor 1 - (n_x/n_0)x^2 - (n_y/n_0)y^2 \rfloor \quad n_x, n_y \ll n_0, \quad (212)$$

then it can be modeled by the FRFT. In fact, the LCT can also model a GRIN system.

Suppose that there is a GRIN medium like the one shown in Figure 2.11. The field distribution at $z = 0$ is $f_0(x, y)$, and the one at $z = L$ is $f_L(x, y)$. If the refractive index of the GRIN medium has a distribution as in Eq. (212), then the relation between $f_0(x, y)$ and $f_L(x, y)$ is

$$f_L(x, y) = \exp\left(-j\frac{2\pi n_0 L}{\lambda} \right) O_{Fx}^{(a_x,b_x,c_x,d_x)} \left(O_{Fy}^{(a_y,b_y,c_y,d_y)}(f_0(x, y)) \right). \quad (213)$$

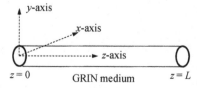

Figure 2.11 A GRIN medium and the directions of the *x*-, *y*-, and *z*-axes.

Here, we use O_{Fx} and O_{Fy} to denote the one-dimensional LCTs along the x-axis and y-axis, respectively, and

$$
\begin{bmatrix} a_x & b_x \\ c_x & d_x \end{bmatrix} = \begin{bmatrix} \cos\phi & w_x\sin\phi \\ \dfrac{\sin\phi}{w_x} & \cos\phi \end{bmatrix}, \begin{bmatrix} a_y & b_y \\ c_y & d_y \end{bmatrix} = \begin{bmatrix} \cos\varphi & w_y\sin\varphi \\ \dfrac{\sin\varphi}{w_y} & \cos\varphi \end{bmatrix},
$$

(214)

where

$$
w_x = \frac{1}{k}\left(\frac{1}{n_x n_0}\right)^{\frac{1}{2}}, \quad w_y = \frac{1}{k}\left(\frac{1}{n_y n_0}\right)^{\frac{1}{2}}, \phi = L\left(\frac{n_x}{n_0}\right)^{\frac{1}{2}}, \varphi = L\left(\frac{n_y}{n_0}\right)^{\frac{1}{2}}. \quad (215)
$$

Note that the effect of the GRIN medium is very similar to the FRFT, except that $\sin\alpha$ is converted into $w_x\sin\alpha$, and that $-\sin\alpha$ is converted into $-\sin\alpha/w_x$.

7.5 Radar System Analysis by the LCT

The LCT can also be used for radar system analysis (Pellat-Finet & Bonnet, 1994). Suppose that there are two spherical disks A and B, as in Figure 2.12. R_A and R_B are the radii of the spherical disks A and B, respectively, and D is the distance between the vertexes of disks A and B.

Suppose that the field distribution on disk A is $F_A(x, y)$. Here, we use the method shown in Figure 2.13 to assign the coordinates (x, y) to the disk. Suppose that R is a point of the disk and that there is a plane perpendicular

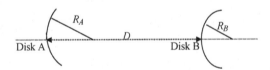

Figure 2.12 A radar system containing two spherical disks and one free space.

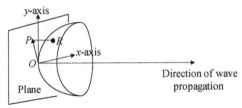

Figure 2.13 A method used to assign the coordinate to the disk.

to the direction of wave propagation and intersecting the disk at O. If the point P is the projection of R on the plane and

$$\overline{OP} = (x, y), \tag{216}$$

then we assign the coordinate of R to be (x, y).

From Pellat-Finet and Bonnet (1994), if the field distribution on the disk B is $F_B(s, h)$, then the relation between $F_A(x, y)$ and $F_B(s, h)$ is

$$F_B(k, h) = e^{-j2\pi D\lambda^{-1}} O_{Sx}^{(R_A, R_B, D)}\left(O_{Sy}^{(R_A, R_B, D)}(F_A(x, y))\right), \tag{217}$$

where the one-dimensional operation $O_{Sx}^{(R_A, R_B, D)}$ is

$$O_{Sx}^{(R_A, R_B, D)}(f(x)) = \sqrt{\frac{j}{\lambda D}}\exp\left(-\frac{j\pi}{\lambda}(R_B^{-1} + D^{-1})s^2\right)$$

$$\times \int_{-\infty}^{\infty} \exp\left(\frac{j2\pi}{\lambda D}sx + \frac{j\pi}{\lambda}(R_A^{-1} - D^{-1})s^2\right)f(x)\,dx \tag{218}$$

and $O_{Sy}^{(R_A, R_B, D)}(f(y))$ has the same form as $O_{Sx}^{(R_A, R_B, D)}(f(x))$. Note that $O_{Sx}^{(R_A, R_B, d)}$ corresponds to the LCT with the following parameters:

$$\begin{bmatrix} a & b \\ c & d \end{bmatrix} = \begin{bmatrix} 1 - R_A^{-1}D & -D/k \\ k(R_A^{-1} - R_B^{-1} + R_A^{-1}R_B^{-1}D) & 1 + R_B^{-1}D \end{bmatrix}. \tag{219}$$

Therefore, with Eqs. (217) and (219), one can also use the LCT to model a radar system.

8. TWO-DIMENSIONAL VERSIONS OF THE LINEAR CANONICAL TRANSFORM

In this section, we introduce the two-dimensional version of the LCT. First, the two-dimensional separable LCT (2-D LCT) is defined as

$$O_{LCT}^{(a_x, b_x, c_x, d_x, a_y, b_y, c_y, d_y)}[g(x, y)] = \left(\frac{1}{j2\pi b_x}\right)^{1/2}\left(\frac{1}{j2\pi b_y}\right)^{1/2}$$

$$\int_{-\infty}^{\infty}\int_{-\infty}^{\infty} \exp\left(\frac{j}{2}\frac{d_x}{b_x}u^2 - j\frac{u}{b_x}x + \frac{j}{2}\frac{a_x}{b_x}x^2\right) \tag{220}$$

$$\exp\left(\frac{j}{2}\frac{d_y}{b_y}v^2 - j\frac{v}{b_y}y + \frac{j}{2}\frac{a_y}{b_y}y^2\right)g(x, y)\,dx\,dy,$$

where

$$a_x d_x - b_x c_x = 1 \quad \text{and} \quad a_y d_y - b_y c_y = 1. \tag{221}$$

The 2-D LCT has eight parameters. It is essentially the direct combination of two one-dimensional LCTs. When $\{a_x, b_x, c_x, d_x, a_y, b_y, c_y, d_y\} = \{\cos\alpha, \sin\alpha, -\sin\alpha, \cos\alpha, \cos\beta, \sin\beta, -\sin\beta, \cos\beta\}$, it reduces to the 2-D FRFT. When $\{a_x, b_x, c_x, d_x, a_y, b_y, c_y, d_y\} = \{\sigma_x^{-1}, 0, 0, \sigma_x, \sigma_y^{-1}, 0, 0, \sigma_y\}$, it reduces to the 2-D scaling operation. When $\{a_x, b_x, c_x, d_x, a_y, b_y, c_y, d_y\} = \{1, \lambda z/2\pi, 0, 1, 1, \lambda z/2\pi, 0, 1\}$, it reduces to the 2-D Fresnel transform.

However, the 2-D LCT is not general enough. To analyze a 2-D optical system more completely, it is proper to use the two-dimensional nonseparable linear canonical transform (2-D NSLCT) (Folland, 1989). In some studies, the 2-D NSLCT is also called the *2-D affine generalized FFT* (Pei & Ding, 2001b), the *2-D Collin's integral* (Collins, 1970), the *2-D nonseparable quadratic phase integral* (Koç, Ozaktas, & Hesselink, 2010), and the *2-D linear canonical transform* (Bastiaans and Alieva, 2007). The definition of the 2-D NSLCT is as follows:

$$\mathbf{O}_{NSLCT}^{(\mathbf{A},\mathbf{B},\mathbf{C},\mathbf{D})}[g(x,y)] = \frac{1}{2\pi[-\det(\mathbf{B})]^{1/2}}$$

$$\times \int_{-\infty}^{\infty} \int_{-\infty}^{\infty} \exp\left[\frac{j}{2\det(\mathbf{B})}\left(k_1 u^2 + k_2 uv + k_3 v^2\right)\right]$$

$$\exp\left\{\frac{j}{\det(\mathbf{B})}\left[(-b_{22}u + b_{12}v)x + (b_{21}u - b_{11}v)y + \frac{1}{2}\left(p_1 x^2 + p_2 xy + p_3 y^2\right)\right]\right\}$$

$$g(x,y)\,dx\,dy, \tag{222}$$

where

$$k_1 = d_{11}b_{22} - d_{12}b_{21}, k_2 = 2(-d_{11}b_{12} + d_{12}b_{11}), k_3 = -d_{21}b_{12} + d_{22}b_{11},$$

$$p_1 = a_{11}b_{22} - a_{21}b_{12}, p_2 = 2(a_{12}b_{22} - a_{22}b_{12}), p_3 = -a_{12}b_{21} + a_{22}b_{11},$$

$$\mathbf{A} = \begin{bmatrix} a_{11} & a_{12} \\ a_{21} & a_{22} \end{bmatrix}, \mathbf{B} = \begin{bmatrix} b_{11} & b_{12} \\ b_{21} & b_{22} \end{bmatrix}, \mathbf{C} = \begin{bmatrix} c_{11} & c_{12} \\ c_{21} & c_{22} \end{bmatrix}, \mathbf{D} = \begin{bmatrix} d_{11} & d_{12} \\ d_{21} & d_{22} \end{bmatrix}. \tag{223}$$

This definition is valid for $\det(\mathbf{B}) \neq 0$. When $\mathbf{B} = 0$, the definition of the 2-D NSLCT is

$$O_{NSLCT}^{(\mathbf{A},\mathbf{B},\mathbf{C},\mathbf{D})}[g(x,y)] =$$

$$[\det(\mathbf{D})]^{1/2}$$

$$\times \exp\left\{\frac{j}{2}\left[(c_{11}d_{11} + c_{12}d_{12})x^2 + 2(c_{11}d_{21} + c_{12}d_{22})xy + (c_{21}d_{21} + c_{22}d_{22})y^2\right]\right\}$$

$$g(d_{11}x + d_{21}y, d_{12}x + d_{22}y)$$

$$(224)$$

The definition of the 2-D NSLCT when $\det(\mathbf{B}) = 0$ but $\mathbf{B} \neq 0$ was shown by Pei and Ding (2001b). Note that from Eq. (223), the 2-D NSLCT has 16 parameters, which should satisfy the following constraints:

$$(1)\ \mathbf{A}^{\mathbf{T}}\mathbf{C} = \mathbf{C}^{\mathbf{T}}\mathbf{A}, (2)\ \mathbf{B}^{\mathbf{T}}\mathbf{D} = \mathbf{D}^{\mathbf{T}}\mathbf{B}, (3)\ \mathbf{A}^{\mathbf{T}}\mathbf{D} - \mathbf{C}^{\mathbf{T}}\mathbf{B} = \mathbf{I} \qquad (225)$$

or

$$(1)\ \mathbf{A}\mathbf{B}^{\mathbf{T}} = \mathbf{B}\mathbf{A}^{\mathbf{T}}, (2)\ \mathbf{C}\mathbf{D}^{\mathbf{T}} = \mathbf{D}\mathbf{C}^{\mathbf{T}}, (3)\ \mathbf{A}\mathbf{D}^{\mathbf{T}} - \mathbf{B}\mathbf{C}^{\mathbf{T}} = \mathbf{I}. \qquad (226)$$

Eqs. (225) and (226) are equivalent. If all of the constraints in Eq. (225) are satisfied, then they are also satisfied in Eq. (226).

The 2-D NSLCT has the following additivity property:

$$O_{NSLCT}^{(\mathbf{P},\mathbf{Q},\mathbf{R},\mathbf{S})}[g(x,y)] = O_{NSLCT}^{(\mathbf{A}',\mathbf{B}',\mathbf{C}',\mathbf{D}')}\left\{O_{NSLCT}^{(\mathbf{A},\mathbf{B},\mathbf{C},\mathbf{D})}[g(x,y)]\right\},$$

$$\text{where } \begin{bmatrix} \mathbf{P} & \mathbf{Q} \\ \mathbf{R} & \mathbf{S} \end{bmatrix} = \begin{bmatrix} \mathbf{A}' & \mathbf{B}' \\ \mathbf{C}' & \mathbf{D}' \end{bmatrix}\begin{bmatrix} \mathbf{A} & \mathbf{B} \\ \mathbf{C} & \mathbf{D} \end{bmatrix}. \qquad (227)$$

That is, the combination of two 2-D NSLCTs can be represented by the product of two matrices.

The 2-D NSLCT is a generalization of many operations. The 2-D separable LCT in Eq. (220) is a special case of the 2-D NSLCT where $a_{12} = a_{21} = b_{12} = b_{21} = c_{12} = c_{21} = d_{12} = d_{21} = 0$. Furthermore, the 2-D FRTT, the 2-D FT, the 2-D inverse FT, the Fresnel transform, the geometric twisting operation, the scaling operation, and the gyrator transform (Rodrigo et al., 2007; Pei & Ding, 2009) are all special cases of the 2-D NSLCT. Figure 2.14 illustrates the relations between the 2-D NSLCT and its special cases.

The 2-D NSLCT is useful for 2-D optical signal analysis (Bastiaans & Alieva, 2007) and 2-D signal processing, especiallyimage processing (Pei & Ding, 2001b). The fast implementation algorithm of the 2-D NSLCT

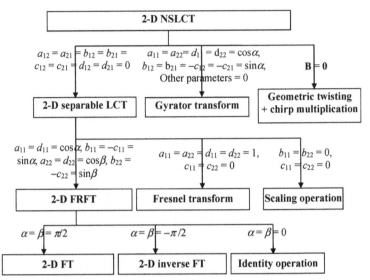

Figure 2.14 The relations among the 2-D NSLCT and its special cases.

was proposed in Koç, Ozaktas, and Hesselink (2010) and Ding, Pei, and Liu (2012). The eigenfunctions of the 2-D NSLCT and the self-imaging phenomena in 2-D optical systems were discussed by Bastiaans and Alieva (2007) and Ding and Pei (2011a).

9. CONCLUSION

In this chapter, we have introduced the definitions, properties, physical meanings, and applications of the LCT and its related operations. Compared to the FT and the FRFT, the LCT is more general and flexible, but its complexity is similar.

The LCT is very useful for signal processing and electromagnetic wave propagation analysis. Most of the applications of the conventional FT can also be treated as the applications of the LCT. Moreover, since the LCT is more flexible than the FT, it can solve many problems that cannot be solved very well by the FT. In many situations, we can use the LCT instead of the FT to achieve even better performance.

REFERENCES

Abe, S., & Sheridan, J. T. (1994a). Generalization of the fractional Fourier transformation to an arbitrary linear lossless transformation: An operator approach. *Journal of Physics A, 27*, 4179–4187.

Abe, S., & Sheridan, J. T. (1994b). Optical operations on wave functions as the Abelian subgroups of the special affine Fourier transformation. *Optics Letters, 19*, 1801–1803.

Alieva, T. (1996). On the self-fractional Fourier functions. *Journal of Physics A—Mathematical and General, 29*, 377–379.

Alieva, T., & Bastiaans, M. J. (2000). On fractional Fourier transform moments. *IEEE Signal Processing Letters, 7*, 320–323.

Alieva, T., & Barbe, A. M. (1997). Self-fractional Fourier functions and selection of modes. *Journal of Physics A—Mathematical and General, 30*, 211–215.

Alieva, T., & Barbe, A. M. (1998). Fractional Fourier and Radon-Wigner transforms of periodic signals. *Signal Processing, 69*, 183–189.

Alieva, T., Lopez, V., Agullo-Lopez, F., & Almeida, L. B. (1994). The fractional Fourier transform in optical propagation problems. *Journal of Modern Optics, 41*(5), 1045–1049.

Almeida, L. B. (1994). The fractional Fourier transform and time-frequency representations. *IEEE Transactions on Signal Processing, 42*, 3084–3091.

Almeida, L. B. (1997). Product and convolution theorems for the fractional Fourier transform. *IEEE Signal Processing Letters, 4*(1), 15–17.

Arikan, O., Kutay, M. A., Ozaktas, H. M., & Akdemir, O. K. (1996). The discrete fractional Fourier transformation. *Proceedings of the IEEE-SP International Symposium on Time-Frequency and Time-Scale Analysis*, 205–207.

Bastiaans, M. J. (1978). The Wigner distribution applied to optical signals and systems. *Optics Communications, 25*, 26–30.

Bastiaans, M. J. (1979). Wigner distribution function and its application to first-order optics. *Journal of the Optical Society of America, 69*, 1710–1716.

Bastiaans, M. J. (1980). Gabor's expansion of a signal into Gaussian elementary signals. *Proceedings of the IEEE, 68*, 594–598.

Bastiaans, M. J. (1989). Propagation laws for the second-order moments of the Wigner distribution function in first-order optical systems. *Optik, 82*, 173–181.

Bastiaans, M. J. (1991). Second-order moments of the Wigner distribution function in first-order optical systems. *Optik, 88*, 163–168.

Bastiaans, M. J., & Alieva, T. (2007). Classification of lossless first-order optical systems and the linear canonical transformation. *Journal of the Optical Society of America A, 24*, 1053–1062.

Bargmann, V. (1961). On a Hilbert space of analytic functions and an associated integral transform, Part I. *Communications on Pure and Applied Mathematics, 14*, 187–214.

Barshan, B., Kutay, M. A., & Ozaktas, H. M. (1997). Optimal filters with linear canonical transformations. *Optics Communications, 135*, 32–36.

Bernardo, L. M. (1996). ABCD matrix formalism of fractional Fourier optics. *Optical Engineering, 35*(3), 732–740.

Bracewell, R. N. (1986). *The Hartley Transform*. New york: Oxford University Press.

Bracewell, R. N. (2000). *The Fourier Transform and Its Applications* (3rd ed.). Boston: McGraw-Hill.

Burrus, C. S. (1977). Index mappings for multidimensional formulation of the DFT and convolution. *IEEE Transactions on Acoustics, Speech, and Signal Processing, 25*, 239–242.

Candan, C. (2007). On higher-order approximations for Hermite-Gaussian functions and discrete fractional Fourier transforms. *IEEE Signal Processing Letters, 14*(10), 699–702.

Caola, M. J. (1991). Self-Fourier functions. *Journal of Physics A: Mathematical and General, 24*, 1143–1144.

Cincotti, G., Gori, F., & Santarsiero, M. (1992). Generalized self-Fourier functions. *Journal of Physics A: Mathematical and General, 25*, 1191–1194.

Classen, T. A. C. M., & Mecklenbrauker, W. F. G. (1980). The Wigner distribution—A tool for time-frequency signal analysis. Part I: Continuous time signals. *Philips Journal of Research, 35*, 217–250.

Collins, S. A. (1970). Lens-system diffraction integral written in terms of matrix optics. *Journal of the Optical Society of America, 60,* 1168–1177.

Condon, E. U. (1937). Immersion of the Fourier transform in a continuous group of functional transformations. *Proceedings of the National Academy of Sciences, 23,* 158–164.

Cong, W. X., Chen, N. X., & Gu, B. Y. (1998). Recursive algorithm for phase retrieval in the fractional Fourier transform domain. *Applied Optics, 37*(29), 6906–6910.

Deng, X., Bihari, B., Gan, J., Zhou, F., & Chen, R. T. (2000). Fast algorithm for chirp transforms with zooming-in ability and its applications. *Journal of the Optical Society of America A, 17*(4), 762–771.

Ding, J. J., & Pei, S. C. (2005). Reducing sampling error by prolate spheroidal wave functions and fractional Fourier transform. *Proceedings of the IEEE International Conference on Acoustics, Speech, and Signal Processing, 4,* 217–220.

Ding, J. J., & Pei, S. C. (2011a). Eigenfunctions and self-imaging phenomena of the two dimensional nonseparable linear canonical transform. *Journal of the Optical Society of America A, 28*(2), 82–95.

Ding, J. J., & Pei, S. C. (2011b). *Discrete linear canonical transform and other additive discrete operations*. Barcelona, Spain: European Signal Processing Conference, 2249–2253.

Ding, J. J., Pei, S. C., & Liu, C. L. (2012). Improved implementation algorithms of the two-dimensional nonseparable linear canonical transform. *Journal of the Optical Society of America A, 29,* 1615–1624.

Erden, M. F., & Ozaktas, H. M. (1998). Synthesis of general linear systems with repeated filtering in consecutive fractional Fourier domains. *Journal of the Optical Society of America A, 15,* 1647–1657.

Flandrin, P. (1988). A time-frequency formulation of optimum detection. *IEEE Transactions on Acoustics, Speech, and Signal Processing, 36,* 1377–1384.

Folland, G. B. (1989). *Harmonic Analysis in Phase Space. The Annals of Math. Studies 122.* Princeton, NJ: Princeton University Press.

Goodman, J. W. (2005). *Introduction to Fourier Optics* (3rd ed.). Roberts & Co., Englewood, CO.

Granieri, S., Arizaga, R., & Sicre, E. E. (1997). Optical correlation based on the fractional Fourier transform. *Applied Optics, 36*(26), 6636–6645.

Guanlei, X., Xiaotong, W., & Xiaogang, X. (2009a). Three uncertainty relations for real signals associated with linear canonical transform. *IET Signal Processing, 3,* 85–92.

Guanlei, X., Xiaotong, W., & Xiaogang, X. (2009b). The logarithmic, Heisenberg's, and short-time uncertainty principles associated with fractional Fourier transform. *Signal Processing, 89,* 339–343.

Guanlei, X., Xiaotong, W., & Xiaogang, X. (2009c). Uncertainty inequalities for linear canonical transform. *IET Signal Processing, 3,* 392–402.

Guanlei, X., Xiaotong, W., & Xiaogang, X. (2009d). Generalized entropic uncertainty principle on fractional Fourier transform. *Signal Processing, 89,* 2692–2697.

Guanlei, X., Xiaotong, W., & Xiaogang, X. (2010). On uncertainty principle for the linear canonical transform of complex signals. *IEEE Transactions on Signal Processing, 58,* 4916–4918.

Healy, J. J., & Sheridan, J. T. (2009). Sampling and discretization of the linear canonical transform. *Signal Processing, 89,* 641–648.

Heisenberg, W. (1927). Über den anschaulichen Inhalt der quantentheoretischen Kinematik und Mechanik. *Zeitschrift für Physik, 4,* 172–198.

Hennelly, B. M., & Sheridan, J. T. (2004). Random phase and jigsaw encryption in the Fresnel domain. *Optical Engineering, 43,* 2239–2249.

Hlawatsch, F., & Boudreaux-Bartels, G. F. (1992). Linear and quadratic time-frequency signal representation. *IEEE Signal Processing Magazine, 9*(2), 21–67.

Hua, J., Liu, L., & Li, G. (1997). Extended fractional Fourier transforms. *Journal of the Optical Society of America A, 14*(12), 3316–3322.

James, D. F. V., & Agarwal, G. S. (1996). The generalized Fresnel transform and its applications to optics. *Optics Communications, 126,* 207–212.

Koç, A., Ozaktas, H. M., & Hesselink, L. (2010). Fast and accurate computation of two-dimensional, non-separable, quadratic-phase integrals. *Journal of the Optical Society of America A, 27,* 1288–1302.

Kutay, M. A., & Ozaktas, H. M. (1998). Optimal image restoration with the fractional Fourier transform. *Journal of the Optical Society of America A, 15*(4), 825–833.

Kutay, M. A., Ozaktas, H. M., Arikan, O., & Onural, L. (1997). Optimal filter in fractional Fourier domains. *IEEE Transactions on Signal Processing, 45*(5), 1129–1143.

Landau, H. J., & Pollak, H. O. (1961). Prolate spheroidal wave functions, Fourier analysis, and uncertainty-II. *Bell System Technical Journal, 40,* 65–84.

Landau, H. J., & Pollak, H. O. (1962). Prolate spheroidal wave functions, Fourier analysis, and uncertainty-III. *Bell System Technical Journal, 41,* 1295–1336.

Liu, S., Yu, L., & Zhu, B. (2001). Optical image encryption by cascaded fractional Fourier transforms with random phase filtering. *Optics Communications, 187,* 57–63.

Lohmann, A. W. (1993). Image rotation, Wigner rotation, and the fractional Fourier transform. *Journal of the Optical Society of America A, 10*(10), 2181–2186.

Lohmann, A. W., Mendlovic, D., & Zalevsky, Z. (1996). Fractional Hilbert transform. *Optics Letters, 21*(4), 281–283.

Lohmann, W., & Soffer, B. H. (1994). Relationship between the Radon-Wigner and the fractional Fourier transform. *Journal of the Optical Society of America A, 11*(6), 1798–1801.

Lohmann, A. W., Zalevsky, Z., & Mendlovic, D. (1996). Synthesis of pattern recognition filters for fractional Fourier processing. *Optics Communications, 128,* 199–204.

Marinho, F. J., & Bernardo, L. M. (1998). Numerical calculation of fractional Fourier transforms with a single fast-Fourier-transform algorithm. *Journal of the Optical Society of America A, 15*(6), 2111–2116.

Martin, W., & Flandrin, P. (1983). Wigner-Ville spectrum analysis of nonstationary processes. *IEEE Transactions on Acoustics, Speech, and Signal Processing, 33*(6), 1461–1470.

McBride, A. C., & Kerr, F. H. (1987). On Namias's fractional Fourier transforms. *IMA Journal of Applied Mathematics, 39,* 159–175.

Mendlovic, D., & Lohmann, A. W. (1997). Space-bandwidth product adaptation and its application to superresolution: Fundamentals. *Journal of the Optical Society of America A, 14*(3), 558–562.

Mendlovic, D., & Ozaktas, H. M. (1993). Fractional Fourier transforms and their optical implementation: I. *Journal of the Optical Society of America A, 10*(9), 1875–1881.

Mendlovic, D., Ozaktas, H. M., & Lohmann, A. W. (1994a). Self-Fourier functions and fractional Fourier transform. *Optics Communications, 105,* 36–38.

Mendlovic, D., Ozaktas, H. M., & Lohmann, A. W. (1994b). Graded-index fibers, Wigner distribution, and the fractional Fourier transform. *Applied Optics, 33,* 6188–6193.

Mendlovic, D., Zalevsky, Z., Lohmann, A. W., & Dorsch, R. G. (1996). Signal spatial-filtering using the localized fractional Fourier transform. *Optics Communications, 126,* 14–18.

Moshinsky, M., & Quesne, C. (1971). Linear canonical transformations and their unitary representations. *Journal of Math and Physics, 12*(8), 1772–1783.

Namias, V. (1980). The fractional order Fourier transform and its application to quantum mechanics. *Journal of the Institute of Mathematics and Its Applications, 25,* 241–265.

Nazarathy, M., & Shamir, J. (1982). First-order optics—A canonical operator representation: Lossless systems. *Journal of the Optical Society of America, 72,* 356–364.

Oppenheim, A. V., & Schafer, R. W. (2010). *Discrete-Time Signal Processing* (3rd ed.). London: Prentice-Hall.

Ozaktas, H. M., & Arikan, O. (1996). Digital computation of the fractional Fourier transform. *IEEE Transactions on Signal Processing, 44*(9), 2141–2150.

Ozaktas, H. M., Barshan, B., Mendlovic, D., & Onural, L. (1994). Convolution, filtering, and multiplexing in fractional Fourier domains and their rotation to chirp and wavelet transform. *Journal of the Optical Society of America A, 11*(2), 547–559.

Ozaktas, H. M., Kutay, M. A., & Mendlovic, D. (1999). Introduction to the fractional Fourier transform and its applications. *Advances in Imaging and Electron Physics, 106*, 239–291.

Ozaktas, H. M., & Mendlovic, D. (1995). Fractional Fourier optics. *Journal of the Optical Society of America A, 12*, 743–751.

Ozaktas, H. M., Zalevsky, Z., & Kutay, M. A. (2000). *The Fractional Fourier Transform with Applications in Optics and Signal Processing*. New York: John Wiley & Sons.

Pei, S. C., & Ding, J. J. (2000a). Closed-form discrete fractional and affine Fourier transforms. *IEEE Transactions on Signal Processing, 48*(5), 1338–1353.

Pei, S. C., & Ding, J. J. (2000b). Simplified fractional Fourier transforms. *Journal of the Optical Society of America A, 17*(12), 2355–2367.

Pei, S. C., & Ding, J. J. (2001a). Relations between the fractional operations and the Wigner distribution, ambiguity function. *IEEE Transactions on Signal Processing, 49*(8), 1638–1655.

Pei, S. C., & Ding, J. J. (2001b). Two-dimensional affine generalized fractional Fourier transform. *IEEE Transactions on Signal Processing, 49*(4), 878–897.

Pei, S. C., & Ding, J. J. (2002a). Eigenfunctions of linear canonical transform. *IEEE Transactions on Signal Processing, 50*(1), 11–26.

Pei, S. C., & Ding, J. J. (2002b). Fractional, canonical, and simplified fractional cosine, sine, and Hartley transforms. *IEEE Transactions on Signal Processing, 50*(7), 1611–1680.

Pei, S. C., & Ding, J. J. (2003a). Eigenfunctions of the offset Fourier, fractional Fourier, and linear canonical transforms. *Journal of the Optical Society of America A, 20*(3), 522–532.

Pei, S. C., & Ding, J. J. (2003b). The generalized radial Hilbert transform and its applications to 2-D edge detection (any direction or specified directions). *International Conference on Acoustics, Speech, and Signal Processing, 2003*(3), 357–360.

Pei, S. C., & Ding, J. J. (2003c). Saving the bandwidth in the fractional domain by generalized Hilbert transform pair relation. *IEEE International Symposium on Circuits and Systems, 4*, 89–92.

Pei, S. C., & Ding, J. J. (2007). Relations between Gabor transforms and fractional Fourier transforms and their applications for signal processing. *IEEE Transactions on Signal Processing, 55*, 4839–4850.

Pei, S. C., & Ding, J. J. (2008). Generalized commuting matrices and their eigenvectors for DFTs, offset DFTs, and other periodic operations. *IEEE Transactions on Signal Processing, 56*, 3891–3904.

Pei, S. C., & Ding, J. J. (2009). Properties, digital implementation, applications, and self image phenomena of the Gyrator transform. *European Signal Processing Conference*, 441–445.

Pei, S. C., & Ding, J. J. (2010). Fractional Fourier transform, Wigner distribution, and filter design for stationary and nonstationary random processes. *IEEE Transactions on Signal Processing, 58*(8), 4079–4092.

Pei, S. C., Hsue, W. L., & Ding, J. J. (2006). Discrete fractional Fourier transform based on new, nearly tridiagonal commuting matrices. *IEEE Transactions on Signal Processing, 54*(10), 3815–3828.

Pei, S. C., Tseng, C. C., Yeh, M. H., & Ding, J. J. (1998). A new definition of continuous fractional Hartley transform. *International Conference on Acoustics, Speech, and Signal Processing, 3*, 1485–1499.

Pei, S. C., & Yeh, M. H. (1997). Improved discrete fractional Fourier transform. *Optics Letters, 22*(14), 1047–1049.

Pei, S. C., Yeh, M. H., & Tseng, C. C. (1999). Discrete fractional Fourier transform based on orthogonal projections. *IEEE Transactions on Signal Processing, 47*(5), 1335–1348.

Pellat-Finet, P., & Bonnet, G. (1994). Fractional order Fourier transform and Fourier optics. *Optics Communications, 111*, 141–154.

Rodrigo, J. A., Alieva, T., & Calvo, M. L. (2007). Gyrator transform: Properties and applications. *Optics Express, 15*, 2190–2203.

Santhanam, B., & McClellan, J. H. (1996). The discrete rotational Fourier transform. *IEEE Transactions on Signal Processing, 42*, 994–998.

Scharf, L. L., & Thomas, J. K. (1998). Wiener filters in canonical coordinates for transform coding, filtering, and quantizing. *IEEE Transactions on Signal Processing, 46*(3), 647–654.

Sharma, K. K., & Joshi, S. D. (2008). Uncertainty principle for real signals in the linear canonical transform domains. *IEEE Transactions on Signal Processing, 56*, 2677–2683.

Shinde, S., & Gadre, V. M. (2001). An uncertainty principle for real signals in the fractional Fourier transform domain. *IEEE Transactions on Signal Processing, 49*, 2545–2548.

Siegman, A. E. (1986). *Lasers.* Mill Valley, Synthesis and Factorization of ABCD Matrices CA: University Science Books. Sec. 20.7 pp. 811–814.

Singh, N., & Sinha, A. (2010). Chaos-based multiple image encryption using multiple canonical transforms. *Optics & Laser Technology, 42*, 724–731.

Slepian, D., & Pollak, H. O. (1961). Prolate spheroidal wave functions, Fourier analysis, and uncertainty-I. *Bell System Technical Journal, 40*, 43–63.

Spiegel, M. R. (2009). *Mathematical Handbook of Formulas and Tables* (3rd ed.). New York: McGraw-Hill.

Stern, A. (2006). Sampling of linear canonical transformed signals. *Signal Processing, 8*, 1421–1425.

Stern, A. (2008). Uncertainty principles in linear canonical transform domain and some of their implications in optics. *Journal of the Optical Society of America A, 25*, 647–652.

Tao, R., Li, B. Z., Wang, Y., & Aggrey, G. K. (2008). On sampling of band-limited signals associated with the linear canonical transform. *IEEE Transactions on Signal Processing, 56*, 5454–5464.

Wiener, N. (1929). Hermitian polynomials and Fourier analysis. *Journal of Mathematics Physics MIT, 18*, 70–73, 1929.

Wigner, E. P. (1932). On the quantum correlation for thermodynamic equilibrium. *Physical Review, 40*, 749–759.

Wolf, K. B. (1979). Canonical transforms. In *Integral Transforms in Science and Engineering.* New York: Plenum Press.

Wolf, K. B., & Krotzsch, G. (1999). Metaxial correction of fractional Fourier transforms. *Journal of the Optical Society of America A, 16*(4), 821–830.

Xia, X. G. (1996). On bandlimited signals with fractional Fourier transform. *IEEE Signal Processing Letters, 3*(3), 72–74.

Yu, L., Huang, M., Wu, L., Lu, Y., Huang, W., Chen, M., & Zhu, Z. (1998). Fractional Fourier transform and the elliptic gradient-index medium. *Optics Communications, 152*, 23–25.

Zalevsky, Z., & Mendlovic, D. (1996). Fractional Wiener filter. *Applied Optics, 35*(20), 3930–3936.

Zalevsky, Z., Mendlovic, D., & Dorsch, R. G. (1996). Gerchberg-Saxton algorithm applied in the fractional Fourier or the Fresnel domain. *Optics Letters, 21*(12), 342–344.

Zayed, A. I. (1996). On the relationship between the Fourier transform and fractional Fourier transform. *IEEE Signal Processing Letters, 3*(12), 310–311.

Zayed, A. I. (1998). Hilbert transform associated with the fractional Fourier transform. *IEEE Signal Processing Letters, 5*(8), 206–208.

Zhao, J., Tao, R., Li, Y. L., & Wang, Y. (2009). Uncertainty principles for linear canonical transform. *IEEE Transactions on Signal Processing, 57*, 2856–2858.

CHAPTER THREE

Mechanical, Electrostatic, and Electromagnetic Manipulation of Microobjects and Nanoobjects in Electron Microscopes

Andrey I. Denisyuk[1], Alexey V. Krasavin[2], Filipp E. Komissarenko[1] and Ivan S. Mukhin[1,3]

[1]St. Petersburg National Research University of Information Technologies, Mechanics, and Optics (ITMO University), 49 Kronverksky, 197101 St. Petersburg, Russia
[2]Department of Physics, King's College London, Strand, London WC2R 2LS, United Kingdom
[3]St. Petersburg Academic University—Nanotechnology Research and Education Centre of the Russian Academy of Sciences, 8/3 Khlopina St., 195220 St. Petersburg, Russia

Contents

1. INTRODUCTION

Nanofabrication methods rely on three basic approaches. One approach, called *top-down*, assumes the fabrication of nanoscale structures from a macroscale object. Typical examples of this method are various

ISSN: 1076-5670
http://dx.doi.org/10.1016/B978-0-12-800264-3.00003-4

lithographic techniques (photolithography, electron beam lithography, focused ion beam etc.). The second approach is called *bottom-up;* here, nano-objects are created by molecular assembly, as in physical and chemical vapor deposition or colloidal chemistry. The third approach is micromanipulation, which is based on the transportation of existing objects (fabricated by other techniques) for the purpose of their investigation, or fabrication of complex structures assembled from basic ones. Micromanipulation techniques can be divided into two groups: manipulation of multiple objects by such tech-niques as electrophoresis, magnetophoresis, and hydrodynamics; and manip-ulation of individual microobjects and nanoobjects. The latter is usually based on various microscopic techniques described below.

Optical (or laser) tweezers, introduced by Ashkin (1970), is supposedly the first method of micromanipulation of individual particles. The underlying physical mechanisms of this phenomenon are now well understood (see, for instance, Neuman & Block, 2004 and references therein). However, the lim-itation of this technique is the size of the transported objects, which should be comparable to the wavelength of light (i.e., not much less than a micrometer).

Other methods for individual particle manipulation rely on using scan-ning probe microscopes. In one of these, the manipulation mechanism is based on mechanical (Sitti & Hashimoto, 2000) or electrostatic (Grobelny *et al.*, 2006; Kim *et al.*, 2011) interaction between the probe and the objects. Also, an object can be glued (Ducker, Senden, & Pashley, 1991) or chem-ically bound (Höppener & Novotny, 2008) to the probe tip. Manipulation in a scanning probe microscope allows for handling particles as small as 5–15 nm in diameter (Kim *et al.*, 2011). However, this kind of manipulation has a limitation: the same probe is used for manipulation and imaging purposes afterwards. So this strategy is not real time, but what is called "move and look." The other shortcomings are that scanning probe microscopes do not allow the visualization of complex structures with high aspect ratios and the modification of the probe during the manipulation process.

The last group of micromanipulation methods is related to manipulation in scanning and transmission electron microscopes (SEMs and TEMs), which is the topic of this chapter. Electron microscopes do not have the lim-itations that optical tweezers or scanning probe microscopes have. Microma-nipulation in SEMs and TEMs allows for handling the individual microobjects and nanoobjects with nanometer precision while the process and the result of the manipulation can be seen in high resolution in real time. In this chapter, we divided the micromanipulation techniques

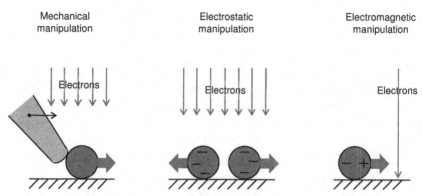

Figure 3.1 Schematic illustration of micromanipulation techniques in electron microscopes: (left) mechanical manipulation (where a micromanipulator is employed, while an electron microscope is used for imaging purposes); (center) electrostatic manipulation (where an electron beam charges particles, which results in attractive or repulsive forces); and (right) electromagnetic manipulation (where an electron beam induces an electromagnetic force exerted on the particle). (See the color plate.)

involving electron microscopes into three main groups (illustrated schematically in Figure 3.1):

- **Mechanical manipulation.** This approach relies on using a number of micromanipulators installed in the chamber of an electron microscope. Here, the manipulation is based on mechanical contact or adhesion interaction between a probe tip of a micromanipulator and an object.

    In addition to this technique, there are noncontact manipulation methods based on intrinsic properties of an electron beam. These are described next

- **Electrostatic manipulation.** Here, the manipulation is based on electrostatic interaction between objects charged under electron beam illumination.

- **Electromagnetic manipulation.** This method is based on the influence of an electromagnetic force created by a beam of passing electrons.

Mechanical manipulation is a well-developed technique, and there is a great many publications in the literature on this topic. Thus, in this chapter, we tried to relate similar approaches, provide references, and focus mostly on several outstanding applications, such as the investigation of mechanical properties of microobjects and nanoobjects. On the other hand, electrostatic and electromagnetic manipulation techniques are far less known. Electrostatic motion of charged objects has been observed in the past; however, guided electrostatic manipulation is not widely used. Electromagnetic manipulation was experimentally demonstrated just a few years ago. The number of publications related to electrostatic and electromagnetic manipulation is very limited. Therefore, here, our review strategy is different: we pay more attention to each publication.

It should be noted that there are also other physical mechanisms of particle manipulation in electron microscopes. For instance, heating of clusters by an electron beam may cause their motion (Williams, 1987) and coalescence (van Huis *et al.*, 2008). Activation of surface bonds by the electron beam also can cause clusters to move (Cretu *et al.*, 2012). Non-Brownian motion of particles was also observed in liquid cells under electron irradiation (Chen & Wen, 2012; Chen *et al.*, 2013). Some studies (e.g., Zheng *et al.*, 2008) describe motion phenomena observed in electron microscopes; however, this motion is not related to a micromanipulator or to the influence of the electron beam. All these experimental results demonstrate a mere particle movement rather than a guided manipulation, and thus these mechanisms will not be taken into account here.

We begin by considering various forces that act on a particle in an electron microscope.

2. OVERVIEW OF FORCES ACTING ON A PARTICLE IN AN ELECTRON MICROSCOPE

2.1 Introduction

Contrary to macroscale, the force of gravity has a very minor impact when the size of an object approaches microscale and nanoscale. At the same time, other forces take over, such as van der Waals forces, external and internal magnetic and electrostatic forces, and capillary forces (Min *et al.*, 2008).

Now, let us consider the forces acting on a particle in electron microscopes. We will examine only a classical arrangement where a particle sits on a substrate, although presently, various liquid cell techniques have become popular in SEMs and TEMs (some aspects of these methods are reviewed in the sections "Electrostatic Manipulation" and "Electromagnetic Manipulation," later in this chapter). We start with the adhesion of a particle to a substrate in vacuum conditions.

2.2 Adhesion of Particles in Vacuum Conditions

The van der Waals attraction between two undeformed spherical particles can be calculated using a simple formula derived by Hamaker (1937):

$$F_{\text{vdW}} = \frac{A}{6} \frac{R_1 R_2}{R_1 + R_2} \frac{1}{z_0^2}, \tag{1}$$

where R_1 and R_2 are the radii of the particles, A is the Hamaker constant for the interacting materials, and z_0 is the separation distance between the particle and the substrate. For most materials, the Hamaker constant is on the order of $10^{-20} - 10^{-19}$ J for interactions in vacuum and about 10 times less than this value for interactions in water. Hamaker constants for some materials can be found in Bergstrom (1997). The separation in the idealized conditions of atomically smooth surface contact is just 3–4 Å. However, the surface roughness of real contacting solids increases the equivalent separation between them (Hu et al., 2010). Here, we calculated the van der Waals attraction force between two spherical particles using Eq. (1) and took $A = 5 \cdot 10^{-20}$ J and $z_0 = 4$ Å. The resulting force is plotted against the particle radius shown in Figure 3.2.

In the case of the deformation of the contacting objects, we need to take into account the contact area, which is assumed to be flat. The attraction force per unit area between two flat surfaces is given by $A/6\pi z_0^3$ (Hamaker,

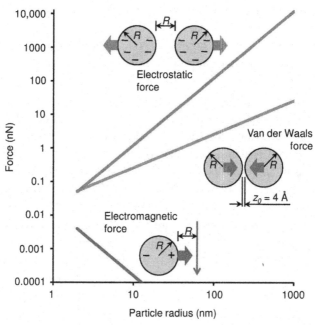

Figure 3.2 Comparison of forces acting on a small spherical particle plotted against a particle radius: van der Waals force between two spheres, Coulomb repulsion force between two charged particles (with separation between them equal to R), and peak electromagnetic force exerted on a particle by a single passing electron (again, with separation equal to R). Forces for very small particles ($R < 2$ nm) are not shown because simplified models used in these calculations would not be valid in such a case. (See the color plate.)

1937). Then the attraction between a deformed spherical particle and a flat substrate (Hu *et al.*, 2010), for instance, can be found as

$$F_{vdW} = \frac{AR}{6z_0^2}\left(1 + \frac{a^2}{Rz_0}\right), \tag{2}$$

where a is the contact radius of the spherical particle after its deformation. Other scenarios for contact mechanics between the deformed solids can be treated using the famous JKR model (Johnson, Kendall, & Roberts, 1971) or DMT model (Derjaguin, Muller, & Toporov, 1975).

This scenario describes the load needed to tear off a particle from a substrate; however, it is also possible to have a sliding motion of a particle over a substrate. The value for the friction force is given by $F_{frict} = \tau \cdot 2\pi a^2$, where a is the contact radius and τ is the interfacial shear strength with a typical value in the range of $10^7–10^9$ Pa (Carpick & Salmeron, 1997). Thus, the friction force in the nanoscale is proportional to the contact area, which differs from the situation in the macroscale.

It should be noted that the way by which the particle was deposited on the substrate can have a crucial influence on the adhesion (Hu et al, 2010). For instance, if the particle was deposited from a liquid suspension by drop-casting and desiccation, then a liquid meniscus is formed between the particle and the flat substrate. It causes a capillary force, which is even stronger than the van der Waals force. The capillary force significantly deforms the particle, and the contact area increases. Being placed in vacuum conditions, the liquid meniscus evaporates and the capillary force disappears; however, the deformation of particles remains, and adhesion in vacuum will be generally the same as it was in air when the capillary force took place (Hu *et al.*, 2010).

In addition, the wet deposition may cause condensation of impurities. Contreras-Naranjo and Ugaz (2013) provided a good illustration of such capillary condensation dynamics (Figure 3.3). Obviously, this also strongly increases the adhesion. Various cases of the contact interface are schematically illustrated in Figure 3.4. As will be shown in the section "Electrostatic Interaction Between Objects Charged due to Electron Irradiation," later in this chapter, micromanipulation systems are able to supply significant forces, which can easily overcome the adhesion of microobjects and nanoobjects.

2.3 Electrostatic Interaction Between Objects Charged due to Electron Irradiation

Charging of dielectric (or ungrounded conductive) objects due to electron bombardment causes electrostatic interaction between them. This can create

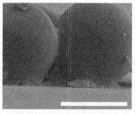

Figure 3.3 SEM images illustrating particle-substrate contact in the cases of dry deposition (left) and wet deposition (center and right) of polystyrene particles on a glass substrate. Wet deposition usually causes the accumulation of impurities underneath the particle. *Reproduced with permission from Contreras-Naranjo and Ugaz (2013). Copyright by Nature Publishing Group.*

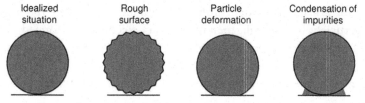

Figure 3.4 Adhesion between a spherical particle and a substrate for various contact interfaces. (See the color plate.)

significant adhesion or repulsion. The physics of charging a dielectric specimen due to electron irradiation is more complex than it looks at first. One can find comprehensive investigations of this phenomenon in Cazaux (2004) or Egerton, Li, and Malac (2004) and references therein.

Scattering of primary electrons during electron-specimen interaction consists of elastic and inelastic collisions. Elastic collision of the primary electrons with specimen atomic nuclei causes deflection of their initial trajectories and formation of backscattered electrons, which are emitted out of the specimen. Ionization of specimen atoms during inelastic collisions leads to the formation of secondary electrons; some of them may leave the specimen during secondary electron emission. Those primary electrons, which lost all their kinetic energy during the collision but did not escape the specimen, become absorbed and charge the specimen negatively. One should note that in the case of low-energy primary electrons, the total charge of the object can be close to zero (or even positive) because the absorbed charge is compensated by higher yield of the secondary electrons. These effects can be illustrated in the following equation:

$$I_b = I_b\eta + I_b\delta + \partial Q/\partial t, \tag{3}$$

where I_b is the current of the primary electron beam, η is the backscattering coefficient, δ is the secondary electron yield, and $\partial Q/\partial t$ is the increase in the accumulated charge. The values for backscattering coefficient and secondary electron yield can be obtained by Monte-Carlo simulation.

However, the entire phenomenon is more complex. The negative charge formed by the trapped electrons can significantly affect the landing energies and trajectories of the primary electrons, as well as the secondary electron emission yield. Thus, the accumulation of the charge is a nonlinear process, and discharge effects also should be taken into account.

Moreover, charging of specimens due to electron irradiation depends not only on specimen electrical properties and primary electron energy, but also on the size and shape of the specimen objects. Let us consider different situations for a dielectric particle sitting on a conductive substrate (Figure 3.5). If the particle is big enough, then the primary electrons are stopped in it, so the particle will be charged negatively. In the case of a small particle, through which primary electrons predominantly pass, the particle will be charged positively because the secondary electron emission will still take place. In

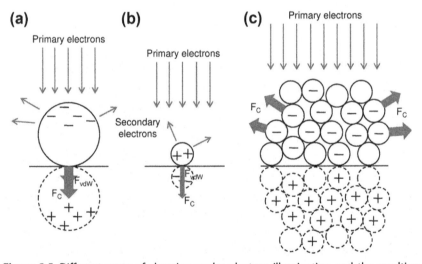

Figure 3.5 Different cases of charging under electron illumination and the resulting adhesion/repulsion for dielectric particles sitting on a conductive substrate ($F_{vdW} =$ van der Waals force; F_C = Coulomb force). (a) A big particle will accumulate a negative charge, which causes increased adhesion to the substrate due to attraction between the charge and its mirror image; (b) a small particle most probably will accumulate a positive charge, and attraction and increased adhesion will take place; (c) small particles piled on top of each other will be charged negatively, which will cause repulsion between them. (See the color plate.)

both cases, the charge accumulated by the particles will form a mirror image with an opposite sign in the substrate, increasing the particle-substrate attraction. However, if small particles are piled on top of each other, then negative charging will take place, which will cause repulsion between the neighboring particles that might be stronger than their adhesion. Some issues regarding the adhesion of micrometer-sized polymer spheres in an SEM were experimentally investigated by Miyazaki et al. (2000a, b), who detected an increase of adhesion for micrometer-sized polymer spheres illuminated by electrons (see the section "Handling micro- and nanoobjects and investigation of their mechanical properties", below in this chapter)

Thus, the charge accumulated by an object under electron beam illumination can take various values and even signs. Now, let us estimate the maximum electrostatic force between two charged spherical particles separated by a distance equal to a particle radius. We assumed that the charge accumulated by each particle would reach the value at which the electric field at the particle surface is 10^9 V/m (higher values would cause discharge due to the field emission effect). The force calculated via Coulomb's law plotted against the particle radius was shown previously in Figure 3.2. One can compare this Coulomb force with the van der Waals force and find that the electrostatic interaction can easily be stronger than adhesion forces, especially for large particles.

2.4 Electromagnetic Force Exerted on a Particle by Fast-Passing Electrons

A fast electron is a source of time-dependent gradient electromagnetic field, possessing a wide range of frequency components in its Fourier spectrum. Flying in the vicinity of a metallic or dielectric particle, this field polarizes it. The resulting, generally multipole, surface charge excitations interact with the source field, exerting a sharp impulse of electromagnetic force as the electron flies by. Thus, if a focused electron beam approaches the particle, the latter experiences a quasi-continuous, time-averaged force, which can be used for its trapping or manipulation.

The problem of momentum transfer to small particles by passing electrons was theoretically investigated by García de Abajo (2004). It was found that the particle momentum change is given by the following frequency integral:

$$\Delta \mathbf{p} = \int_0^\infty \mathbf{F}(\omega)d\omega, \tag{4}$$

where

$$\mathbf{F}(\omega) = \frac{1}{4\pi^2} Re\Bigg\{ \oint_S d\mathbf{s}(\mathbf{E}(\mathbf{s},\omega)(\mathbf{E}(\mathbf{s},\omega)\cdot\widehat{\mathbf{n}})^* + \mathbf{H}(\mathbf{s},\omega)$$
$$\times (\mathbf{H}(\mathbf{s},\omega)^*\cdot\widehat{\mathbf{n}})^* - \frac{\widehat{\mathbf{n}}}{2}\left(|\mathbf{E}(\mathbf{s},\omega)|^2 + |\mathbf{H}(\mathbf{s},\omega)|^2\right)\Bigg)\Bigg\}$$

(5)

defines the relative input of the Fourier components of the total electro-magnetic field, including an external field (produced by the electron) and an induced field (due to multipole excitation in the particle). Here, the integration is done over the nanoparticle surface, and $\widehat{\mathbf{n}}$ is the surface normal. The total field can be found expressing the source field in the basis of spherical functions with an origin at the nanoparticle center and matching them with the induced multipole fields via standard boundary conditions. The derived formalism was applied to both dielectric (Al_2O_3) and metallic (Ag) nanoparticles. In both nanoparticle types, and for all the distances between the center of the particle and the beam (impact parameters), the longitudinal component of the momentum supplied to the particle, parallel to the velocity of the electrons, always points in the same direction as the latter. This means that the particle is pushed along the direction of the beam. The magnitude of the supplied momentum in this direction (and therefore the force experienced by the particle) drastically increases as the impact parameter becomes smaller.

The situation is more interesting, though, for the transverse component of the supplied momentum, perpendicular to the beam direction. For the alumina nanoparticle, for all the impact parameters, this component is directed toward the beam. It also rapidly increases with the decrease of the impact parameter and tends toward zero as the parameter decreases. Interestingly, the latter happens at a much slower rate than for the longitudinal component; therefore, at larger impact parameters, the transverse force component dominates. For the smallest particle considered ($r = 10$ nm), the monotonic increase of this momentum component at small impact parameters (~ 50 nm) is perturbed. For metallic nanoparticles, however, in the analogous situation, the transverse force even changes sign. So one can conclude that while at large impact parameters, the metallic particle is attracted to the electron beam, at small impact parameters, the particle is actually pushed away from it. The domination of the transverse component at large impact parameters happens for metallic particles as well. At the same time, the ratio between the components in the limit of nanometer impact parameters depends on the particle type and size [c.f. the data from García de Abajo (2004) and Reyes-Coronado *et al.* (2010)]. In addition, for both particles, the passing electron induces a torque on a particle that makes it rotate.

A comprehensive review of the physics of force formation in the case of metallic nanoparticles was given by Reyes-Coronado *et al.* (2010). In particular, the mechanism reveals itself when the dielectric response function of the material is defined by the Drude model (parameters corresponding to aluminium were considered). As previously noted, the electromagnetic field of a fast (> 100 keV) electron possesses in its Fourier spectrum the whole range of frequencies from 0 to tens of electronvolts, matching the energies of the plasmonic multipole resonances in the nanoparticle. Therefore, certain frequency components of this spectrum resonantly excite the latter, which results in an inevitable transfer of momentum from the electron to the particle. The higher-frequency components are present mostly in the vicinity of the electron trajectory (since they have more rapid evanescent decay), while the lower-frequency components penetrate into space for greater distances. Through this, the efficiency of the multipole excitation (higher multipoles have higher resonant frequencies), and therefore the overall momentum transfer, depend on the impact parameter.

For a small ($r = 1$ nm) particle, for all the exited plasmonic resonances, the longitudinal component of the force is pointed along the beam direction. The direction of the transverse force, though, exhibits more interesting behavior. If the frequency component of the fast electron electromagnetic field is below the multipole resonant frequency, then this multipole's impact is attractive (the excited multipole is "in phase" with the source field). On the other hand, if the frequency component is above the resonance, this multipole's impact is repulsive (the multipole is "out of phase") (Figure 3.6). For bigger impact parameters, and therefore dominating low frequency components in the Fourier spectrum of the source electromagnetic field, the following is true:

1. The excited dipole resonance prevails over higher-frequency multipole resonances.
2. The frequency components of the source field are far below the dipole resonant frequency. This results in the "dipole in phase" situation and the attractive transverse force.

For smaller impact parameters (< 5 nm), the repulsive force frequency components become dominant, resulting in a repulsive total transverse force

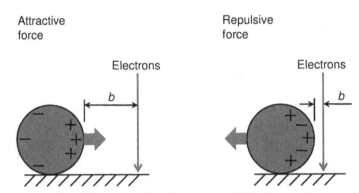

Figure 3.6 Schematic illustration of an electromagnetic force exerted on a particle by a fast-passing electron. The force can be attractive or repulsive, depending on the impact parameter b (following Batson *et al.*, 2011). (See the color plate.)

exerted on the particle by the beam, which explains the behavior observed in García de Abajo (2004). This phenomenon is due to both the multipolar nature of excitation and the retardation effects.

For small impact parameters, these effects are even more pronounced for larger ($r = 40$ nm) particles, where the higher-order plasmonic multipoles are excited more efficiently and the retardation effects are more pronounced. This results in a much larger repulsive force (bigger than a possible attractive one at larger impact parameters), and even in substantial anti-pushing longitudinal frequency components, the total longitudinal force still remains along the direction of the beam, but it does not rise monotonically. Rather, it starts to decrease for ~ 1-nm impact parameters. Compared with the $r = 10$ nm particle, the longitudinal component of the force is larger by one to three orders (depending in the impact parameter), while the transverse component is larger by about one order. Finally, the interaction of the beam with a gold nanoparticle with tabulated frequency dependence of the dielectric constant was considered and found significantly different from the one given by the Drude model; this was due to interband transition contributions. It was determined that the force exerted on the particle in this case is defined by high (> 10 eV) frequency tails of the total electromagnetic field, so in this case, it is more proper to speak about the dielectric rather than the plasmonic nature of the force. At the same time, it was noted that the qualitative dependence of the force's spectral components on the impact parameter remain the same as in the plasmonic case, its value becoming approximately one to two orders of magnitude larger.

Now, let us try to estimate the order of the electromagnetic force produced by a single electron using a very simple approach. The transverse component of the electric field produced by a relativistic electron at distance r is given by

$$E = \frac{1}{4\pi\varepsilon_0} \frac{e}{r^2} \frac{1}{\sqrt{1 - \left(\frac{v}{c}\right)^2}}. \tag{6}$$

This electric field induces polarization of the particle; thus, the force acting on the particle is obtained as

$$\mathbf{F} = \int \mathbf{P} \cdot \nabla \mathbf{E} \cdot dV. \tag{7}$$

For the case of a spherical particle with radius R and relative permittivity ε, we can estimate this force as

$$F = 4\pi R^3 \varepsilon_0 \frac{\varepsilon - 1}{\varepsilon + 2} \cdot E \cdot \nabla E. \tag{8}$$

The calculated electromagnetic force for the case of electron velocity $v/c = 0.7$ and particle relative permittivity $\varepsilon = 3$, plotted aganst the particle radius, was shown previously in Figure 3.2. Separation between the electron trajectory and the particle (impact parameter) is considered to be equal to the radius of the particle.

One can notice that the electromagnetic force produced by a single passing electron is very small and can hardly be stronger than the van der Waals adhesion force, even for particles that are only a few nanometers in size. Thus, a passing electron with its field cannot tear off a particle from a substrate. It can only induce the dragging of a very small cluster over a surface or the motion of bigger particles in a liquid environment (see the section "Electromagnetic Manipulation," later in this chapter). Let us also estimate momentum transfer induced by a passing electron and time-averaged force produced by an electron beam using Eqs. (6)–(8). For instance, for the case of a particle with $R = 10$ nm and an electron traveling at $v/c = 0.7$ at the distance of 10 nm from the particle surface, we calculate the peak force as $F \approx 0.15$ pN. So the momentum transfer is estimated as $p \approx 8 \times 10^{-30}$ N s ($p \approx F \times \tau$, where $\tau \approx 50 \times 10^{-18}$ s is the dwell time of an electron near the particle). The time-averaged force induced by a 1-nA electron beam (to cite one example) would be $F_{av} \approx 5 \times 10^{-20}$ N ($F_{av} \approx p/t$, where $t = 0.16$ ns is an average time interval between neighboring electrons in the beam).

We would like to note that our estimated value of momentum transfer has the same order as the exact value obtained by García de Abajo (2004) for Al_2O_3 particles of the same size and passing electrons of the same parameters.

3. MECHANICAL MANIPULATION

3.1 Introduction

Mechanical manipulation of microobjects and nanoobjects relies on the usage of various micromanipulators with attached needle-shaped probe tips or microgrippers. All these devices are either commercial or self-made. This manipulation technique is usually based on the mechanical pushing of an object with a probe tip. Pick-and-place manipulation is also possible, either using a microgripper or a single probe tip but involving adhesion mechanisms (Figure 3.7).

The first work that describes the concept and implementation of a manipulation system installed in an electron microscope chamber was published by Hatamura and Morishita (1990). According to their concept, an operator monitors the process on a magnified three-dimensional (3-D) image from a stereo SEM and manipulates two nanorobots in the vacuum chamber via a bilateral joystick mechanism. They reported that using this system, the operator was able to control the position of micrometer-sized objects with an accuracy of 10 nm. A prototype of a three-axis nanorobot manipulator was developed and installed in a chamber of a stereo SEM that was equipped with a specimen stage that can move along one horizontal axis. A three-axis nanorobot was based on piezoelectric actuators, while the system was additionally equipped with a single-axis force sensor.

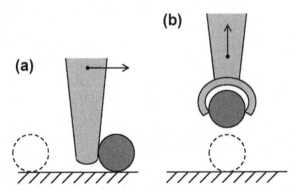

Figure 3.7 Schematic illustration of the mechanical pushing of an object with (a) a needle-shaped probe tip and (b) pick-and-place manipulation using a microgripper. (See the color plate.)

Hatamura and Morishita (1990) tested this prototype by making submicrometer scratches on an aluminium substrate. Miyazaki and Sato (1996) performed the first research on manipulation of microobjects in an SEM. The implemented technique was based on the use of a special micromanipulator with a nominal accuracy of 10 nm installed in the specimen chamber. Using this instrument, polymer microspheres were assembled in ordered structures on a substrate by pick-and-place manipulation using a needle-shaped probe tip and employing adhesion via van der Waals forces.

Modern micromanipulation systems for SEM/TEM are described by Fukuda, Arai, and Nakajima (2013) and references therein. Therefore, this section of this chapter is mainly dedicated to applying micromanipulation systems for handling microobjects and nanoobjects and investigating their mechanical properties. In addition, the principles and characteristics of systems are described briefly.

Some basic principles of modern manipulation techniques in SEMs were reviewed by Jasper (2011). The main part of a micromanipulator is an actuator that should be able to move with nanoscale accuracy. In order to implement such precise movements, the actuator exploits some physical effects. In particular, actuators can be piezoelectric, electrostatic, thermal, or magnetostrictive. Piezoelectric actuators, which are based on the reverse piezoelectric effect, are the most common.

Piezoelectric actuators allow smooth and precise motion. However, their working range is very limited, normally being just a few micrometers. In order to perform coarse positioning in wide range accompanied by fine positioning in short range, a stick–slip principle can be employed based on the inertia and friction. As illustrated in Figure 3.8, a moving part is connected to a fixed piezoelectric actuator using a friction contact. Thus, in the case of slow deformation of the actuator, the moving part follows it (the stick phase). In the case of abrupt contraction of the actuator, the moving part slips and does not move (the slip phase). The sequence, consisting of alternating stick and slip motion phases, provides coarse positioning over a wide range, while the stick phase itself provides fine positioning over a short range.

3.2 Commercial Micromanipulators

Micromanipulators for electron microscopes are produced by several companies. The leading manufacturers include Kleindiek Nanotechnik (based in Reutlingen, Germany), Klocke Nanotechnick (Aachen, Germany), Oxford Instruments (Abingdon, England), Zyvex Instruments (Richardson, Texas), SmarAct (Oldenburg, Germany), FEI (Hillsboro, Oregon), Imina

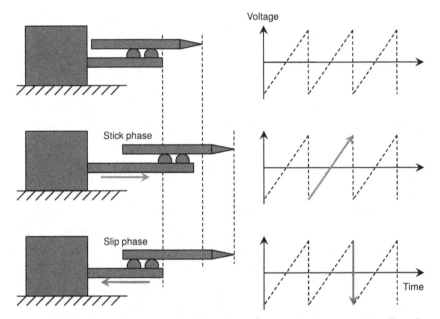

Figure 3.8 The stick-slip principle of motion of piezoelectric actuators. (See the color plate.)

Technologies (Lausanne, Switzerland), Hummingbird Scientific (Lacey, Washington), and Xidex (Austin, Texas).

These manipulators differ in dimensions; some of them are compact and can be mounted inside the SEM chamber on a wall or on a translation stage, while others have to be port-mounted on the wall of the chamber partly outside. Also, a micromanipulator made by Imina Technologies is a mobile robot called miBot, while Hummingbird Scientific offers a special holder for manipulation in TEMs. All manipulators have low vibration constructions to achieve high resolution, low backlash, and drift.

In all these micromanipulators, the piezoelectric effect is employed for at least fine motion. Coarse motion is achieved either by the stick-slip principle or a separate motorized control. Almost all manipulators have a Cartesian coordinate system, but a polar coordinate system and a combination of the Cartesian and the polar systems are also used. A micromanipulator OmniProbe from Oxford Instruments provides the Cartesian system with additional rotational motion. Also, the possibility of multiple degrees of freedom is provided by SL-line and SR-line micromanipulators from SmarAct.

The described micromanipulators provide precision of motion ranging from a few nanometers down to subnanometers. Some micromanipulators have a position sensor that provides a feedback loop. Micromanipulators can also be equipped with various optional tools, such as microgrippers and force sensors. Special probe tips suitable for mechanical and electrical characterization also can be installed. Micromanipulators are able to achieve significant forces, which are sufficient for handling any kind of microobjects and nanoobjects. For instance, an MM3A-EM micromanipulator from Kleindiek Nanotechnik provides a holding force of 1 N, whereas the gripping force ranges from 5 to 5000 μN. The main characteristics of the reviewed micromanipulators are summarized in Table 3.1.

3.3 Handling Microobjects and Nanoobjects and Investigation of Their Mechanical Properties

A standard application of manipulators in electron microscopy (particularly, in dual electron-ion beam workstations) is the fabrication of TEM lamellae. The basic procedure of the fabrication and some of its features are described in various publications (e.g., Mayer et al., 2007). In this procedure, the lamella is cut by focused ion beam milling, whereas manipulation is used to transport a probe tip with a welded lamella to a TEM grid.

More advanced manipulation of microobjects and nanoobjects requires either the usage of modified commercial micromanipulators or developing self-made manipulators. For instance, Peng et al. (2004) demonstrated the gripping of nanoobjects using a manipulation system comprised from four Kleindiek Nanotechnik nanoprobes installed in an SEM. A number of studies are dedicated to the manipulation and characterization of carbon nanotubes in SEMs using manipulation systems with multiple degrees of freedom. For example, Yu et al. (1999) developed a piezomanipulator with XYZ translation and one rotational motion, whereas Fukuda et al. (2003) constructed a nanorobotic manipulator with 16 degrees of freedom. Hänel et al. (2006) used a scanning tunneling microscope integrated in an SEM for the manipulation of organic nanocrystallites. Zhang et al. (2013) developed a nanomanipulation system that can be installed in the SEM chamber via a load-lock. Micromanipulators also can be used in TEMs, although the space there is limited. For instance, Dong et al. (2008) developed a system that is integrated into a TEM holder.

Micromanipulators can use not only needle-shaped probe tips, but also microgrippers. The gripping mechanism can be based on physical effects of varying types, such as piezoelectric (Clévy et al., 2005), electrostatic (Mølhave et al., 2006), electrothermal (Cagliani et al., 2010), or a shape memory effect (Nakazato et al., 2009). The most advanced microgripper,

Table 3.1 Main Characteristics of Commercial Micromanipulators

Manufacturer	Mounting	Motion Principle	Coordinate System	Motion Precision	Optional Tools
Kleindiek Nanotechnik (www.nanotechnik.com)	Inside the SEM chamber	Piezo, stick–slip; holding force: 1 N	Polar + optional rotation	0.5–5 nm	Microgripper (gripping force: 5–5000 µN), rotational tip, force sensor, electrical characterization
Klocke Nanotechnik (www.nanomotor.de)	Inside SEM chamber	Piezo	Cartesian	1 nm	Position sensor, force sensor
Oxford Instruments (www.oxford-instruments.com/)	Flange-mounted, partly outside SEM	Piezo	Cartesian and optional rotation	Sub-nm, 10 nm in feedback loop	Position sensor, electrical characterization
Zyvex (www.zyvex.com)	Inside SEM chamber	–	Cartesian	5 nm	
SmarAct (www.smaract.de)	Inside SEM chamber	Piezo, stick–slip	Cartesian, optional multiple degrees of freedom	Sub-nm, 1 nm in feedback loop	Microgripper
Imina Technologies (www.imina.ch)	Mobile	Piezo	Cartesian and polar	0.5 nm	Microgripper, optical fiber
Hummingbird Scientific (hummingbirdscientific.com)	TEM mounted	Fine piezo and motorized coarse	Cartesian	1 nm	N/A
Xidex (www.xidex.com/)	Inside SEM chamber	Piezo, stick–slip	Cartesian	Sub-nm	Microgripper, force sensor

which was tested by Cagliani *et al.* (2010), was able to handle nanowires and nanotubes of sub-100-nm diameters.

This material has described mechanical micromanipulations in electron microscopes performed manually when an operator controls the movement, monitoring it via SEM/TEM images. However, presently this process can be done automatically, with the system performing all the manipulations itself via vision-based motion control. Such automatic manipulation is faster, and it does not depend on the operator's skills. The first automatic manipulation system for an SEM was constructed by Kasaya *et al.* (2004), who demonstrated pick-and-place manipulation of 30-μm metal spheres. The motion of the probe tip relied on an image recognition system that used filtered SEM images to detect edge fragments. Moreover, a force sensor gave information about this event. Pickup of a sphere by a probe tip occurred via the adhesion mechanism (further investigation of adhesion in SEMs performed by the same group is described later in this chapter). Some newer research on this topic (Eichhorn *et al.*, 2009) describes the use of a microgripper in the automatic mode for handling carbon nanotubes. Other studies by the same scientists (Jasper & Fatikow, 2010; Jasper, 2011) also investigated ways of overcoming a natural limitation of an SEM, which reduces the speed of manipulations: according to these findings, scanned images of a moving object obtained in an SEM give a distorted picture. Increasing the scanning speed reduces the distortion but causes noisy, poor-quality images. A possible solution, which allows tracking the object fast and precisely, is a technique involving two line scans, which is schematically illustrated in Figure 3.9.

A number of other studies have been dedicated to the visual tracking of motion in an automatic mode, and they have looked at techniques such as improving electron image processing using contour models (Ru *et al.*, 2012a) or contact detection purely based on image recognition (Ru & To,

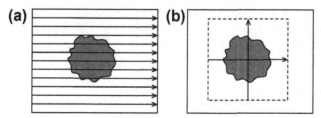

Figure 3.9 Schematic illustration of tracking of an object by (a) normal SEM imaging and (b) a two-line scan. The latter is more suitable for automatic tracking. (See the color plate.)

2012b). Visual contact detection was also employed by Zhang *et al.* (2012) during the fabrication of nanowire-based field-effect transistors. The research performed by Ye *et al.* (2013) is dedicated to three-dimensional (3-D) automatic navigation, which is needed to pick up a freestanding nanowire with a probe tip [see also Ye (2012) and references therein].

Another classical application of micromanipulators is providing contacts to local positions on the surface of microobjects and nanoobjects, which are mostly one-dimensional (1-D) structures (e.g., nanowires, nanotubes, and nanobelts) to investigate their electrical characteristics (the electron beam is usually blanked for the time of the measurement). While this technique does not require any manipulation with objects, there are studies where manipulation ability has been employed. Manipulation can be used for either mechanical deformation of the object in order to study dependence of electrical characteristics on the deformation, or for assembling of objects in order to study characteristics of complex structures. For instance, Kim *et al.* (2003) studied electrical characteristics of multiwall carbon nanotubes in an SEM, equipped with two micromanipulators with tungsten probe tips. In particular, the authors investigated the field emission properties of a single nanotube. The nanotube was welded to a tip of the first manipulator, whereas the second manipulator was positioned a few micrometers from a freestanding end of the nanotube. Another section of the same paper was dedicated to studying the current-voltage characteristics of the nanotube bent between two tips of the micromanipulators. Bussolotti *et al.* (2007) also investigated the electrical characteristics of multiwall nanotubes with the help of micromanipulators. The nanotubes were grown on a nickel surface; the approaching tip of a micromanipulator was welded to a freestanding end of the nanotube to provide an electrical contact. Measurements of current-voltage characteristics showed that the electrical resistance of the nanotube increases several orders of magnitude upon bending the tube from a straight shape to a U-shape. Liu *et al.* (2008) investigated the current-voltage characteristics of a bent ZnO nanowire. The experiment was performed in a TEM equipped with a special specimen holder with a piezomanipulator. Using the high-resolution TEM technique helped to determine a crystalline structure of the bent nanowire, which was linked to its electrical properties (Figure 3.10). Also, a recent study by Mai *et al.* (2012) is dedicated to the mechanical and electrical characterization of ZnO nanorings. The authors studied not only the current-voltage characteristics of a nanoring, but also its mechanical properties, by compressing an individual nanoring with

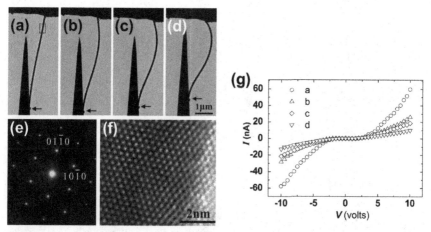

Figure 3.10 Measurement of current-voltage characteristics of a ZnO nanowire upon its bending using a piezomanipulator integrated into a specimen holder of a TEM. *Reproduced with permission from Liu et al. (2008). Copyright by AIP Publishing LLC.*

two micromanipulators in an SEM. Other research related to mechanical characterization is reviewed in detail next.

Investigation of various mechanical properties of microobjects and nano-objects (e.g., adhesion, hardness, and friction) is another application of micromanipulators in electron microscopes. Most of the research in the field of quantitative nanomechanics was performed with the help of atomic force microscopy (e.g. Götzinger & Peukert, 2003). But only a micromanipulator with a force sensor installed in an electron microscope provides a unique capability of measurement and real-time visualization of the process. However, as already noted, an electron beam charges objects, which can greatly affect the measurements. Nevertheless, the analysis of the experimental results obtained at these conditions provides a guideline for reliable micromanipulation, particularly in automatic mode.

For instance, some experimental aspects of adhesion of polymer micrometer-sized spheres under electron beam illumination were investigated by Miyazaki *et al.* (2000a, b), who developed a special adhesion force measurement system installed in an SEM. This system was based on a movable metalized probe tip and a substrate with laser interferometer to detect the displacement. This allowed the measurement of an adhesion force of micrometer-sized polymer particles deposited on a gold substrate. The obtained value for adhesion ranged from 50 to 3000 nN. In particular, the authors observed an increase of the adhesion force with the time spent

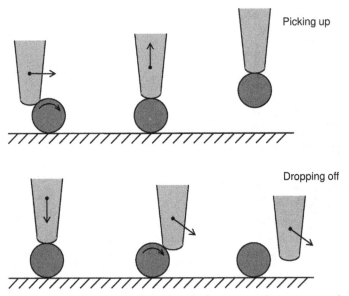

Picking up

Dropping off

Figure 3.11 Schematic illustration of a pick-and-place manipulation strategy based on an adhesion effect (following Saito *et al.*, 2002). (See the color plate.)

on electron beam illumination, which was attributed to charging and an electrostatic contribution to adhesion. Some further experiments on adhesion measurement and adhesion-based manipulation were performed by Saito *et al.* (2002) using a needle-shaped probe tip in an SEM. Here, the kinematics of the rolling and slipping motion of microspheres was investigated both theoretically and experimentally. As a result, a reliable method of pick-and-place operations was proposed (as shown in Figure 3.11).

Mechanical micromanipulation also can be employed to investigate the hardness of nanoobjects. For instance, Enomoto *et al.* (2006) experimentally measured Young's modulus of carbon nanotubes fabricated using different methods: arc discharge, catalytic chemical vapor deposition, and thermal chemical vapor deposition. The apparatus for Young's modulus measurement was installed in a TEM and consisted from a stationary stage with a specimen and an XYZ piezodriving stage with an atomic force microscope cantilever attached. The measurement principle was based on bending the nanotube upon an applied force (Figure 3.12). Moreover, TEM operating at 200 kV enabled the visualization of a crystalline structure of the nanotube. The authors demonstrated that nanotubes obtained by arc discharge have the highest crystallinity, and their measured Young's modulus of 3.3 TPa is very close to the theoretical value. Nakajima, Arai, and Fukuda (2006) developed

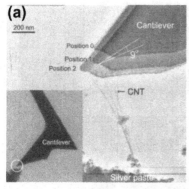
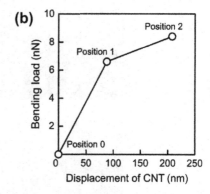

Figure 3.12 Experimental investigation of the Young's modulus of a carbon nanotube using a special apparatus installed in a TEM. *Reproduced with permission from Enomoto et al. (2006). Copyright by AIP Publishing LLC.*

a hybrid manipulation system with multiple degrees of freedom, which can be installed in a TEM or an SEM. A manipulator with 8 degrees of freedom placed inside an SEM performed preliminary positioning. Then a small unit with 3 degrees of freedom could be fitted in a limited volume of a TEM for final manipulation with a higher precision. To demonstrate the effectiveness of the system, the measurement of the Young's modulus of carbon nano-tubes was performed. Investigation of mechanical properties of another one-dimensional nanostructure, a silver nanowire, was performed by Vlas-sov *et al.* (2014). The experiments were conducted inside an SEM equipped with a self-made force sensor. The Young's modulus and yield strength were found to be 90 GPa and 4.8 GPa, respectively. Notably, no dependence on the nanowire diameter was observed. High fatigue resistance of silver nano-wires was also demonstrated.

Measurement of friction at the nanoscale can also be done with the help of mechanical micromanipulation in electron microscopes. A series of recent experiments (Vlassov *et al.*, 2011; Polyakov *et al.*, 2011, 2012, 2014) presents results of investigations of the tribological properties of various nanoobjects. For these experiments, a micromanipulation system was developed and installed in an SEM. The system consisted of an atomic force microscope cantilever glued to a quartz tuning fork force sensor. The authors studied the static friction of 150-nm gold nanoparticles and found it to be in the range of 40–750 nN. ZnO nanowires were another object of investigation, where translation of a nanowire over a surface and the resulting elastic de-formations were used to determine a distributed friction force. Also, the

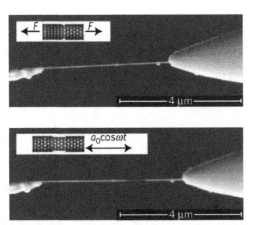

Figure 3.13 Experimental investigation of interlayer friction in a boron nitride nano-tube performed in an SEM. *Reproduced with permission from Niguès et al. (2014). Copyright by Nature Publishing Group.* (See the color plate.)

dependence of static friction of CuO nanowires on the surface roughness was investigated. The experimental results suggest a considerably smaller friction force for smoother surfaces; this differs from the macroscale situation, where friction has no dependence on the contact area. The same group carried out another study that looked at the tribological properties of silver nanodumbbells. These objects are characterized by reduced contact area and adhesion in comparison to nanowires. Different types of nanodumbbell motion (i.e., rolling, sliding, and rotation) were demonstrated. Another interesting investigation of the friction of nanomaterials was performed by Niguès *et al.* (2014). They studied interlayer friction in a boron nitride nanotube using an SEM and used a quartz tuning fork as a force sensor. A nanotube was torn apart, and the measured friction was found to be proportional to the contact area (see Figure 3.13).

4. ELECTROSTATIC MANIPULATION

The electrostatic manipulation technique is based on the electrostatic interaction between objects charged under electron beam illumination. Random motion or deformation of the specimen objects due to charging is an unwanted effect, which sometimes can be observed in an electron microscope. This includes for example, the accidental shifting of dielectric particles during SEM imaging or tearing a thin film in the case of a TEM. However, the effect of specimen charging can also be employed to produce controllable motion (i.e., manipulation).

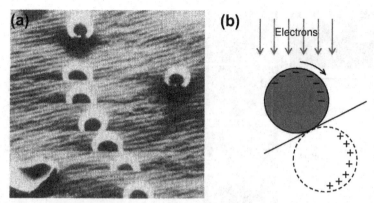

Figure 3.14 (a) Rolling motion of a spherical polystyrene particle observed on a single SEM image with a 40-s recording rate. Reproduced with permission from Krakow and Nixon (1977). Copyright by IEEE. (b) Schematic illustration of an uphill motion mechanism that occurs due to a torque that is produced by an attractive force between particle charge and its mirror image in a substrate when the latter is tilted (following Krakow & Nixon, 1977). (See the color plate.)

One of the first works related to this topic was published by Krakow and Nixon (1977). The authors studied charging phenomenon during SEM imaging of 10-μm polystyrene and glass spherical particles mounted on a gold-coated grating substrate. Total charge accumulated by a particle was found to be 10^{-12} C, which was estimated from the distortion of the SEM image. When the substrate was tilted more than 17°, directed rolling motion of particles was observed (Figure 3.14a). Moreover, when a low-energy electron beam (10 kV) was applied, the spherical particles rolled uphill, whereas downhill motion was observed at higher electron energies (i.e., 20 kV). This effect occurred due to the torque produced by an attractive force between the negative charge in the particle and an induced mirror charge in the conductive substrate. This explains the uphill motion at 10 kV that is illustrated in Figure 3.14b. Downhill motion at 20 kV is believed to be due to deeper localization of the negative charge within the particle. As a result, the torque appeared to be the opposite. Krakow and Nixon also performed a theoretical study of the effect to investigate the qualitative dependence of the torque and velocity of a sphere rolling, as opposed to the substrate tilt angle.

Another study is dedicated to the controllable deflection of Si nanowires in response to electron beam illumination (Fukata *et al.*, 2005). The authors observed motion while imaging a pair of nanowires grown on a Si wafer. They found that the separation between the nanowires increases upon the

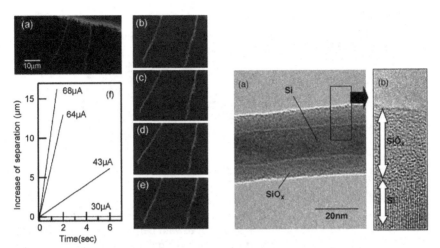

Figure 3.15 SEM images showing controllable deflection of Si nanowires due to their charging under electron beam illumination. The increase of separation between the nanowires is plotted against the time of illumination for different electron beam currents. (left). TEM images confirm the structure of a Si nanowire with a SiO$_x$ surface layer, which is responsible for charge accumulation (right). *Reproduced with permission from from Fukata* et al. *(2005). Copyright by IOP Publishing.* (See the color plate.)

increase of the electron beam current and imaging time (Figure 3.15). After electron beam illumination finishes, the distance between the nanowires returns to the initial value immediately. The authors suggest that the motion was caused by the Coulomb repulsive interaction between the nanowires. The charge is believed to be accumulated in a SiO$_x$ surface layer, which covers the Si crystalline core of the nanowire. The conductive Si core is also responsible for the fast discharge after the completion of electron beam illumination.

Charge patterns created on a dielectric substrate in an SEM can be used for nanoparticle assembly, as demonstrated by Zonnevylle *et al.* (2009). Positively charged Pd nanoparticles were created in an Ar atmosphere from a glowing wire generator in a deposition chamber. Gas flow with suspended nanoparticles passes through a differential mobility analyzer for the selection of particles of a certain size and charge. At the same time, in an SEM chamber, Si substrate with a 6-μm-thick layer of Si$_3$N$_4$ was exposed by a 6-kV electron beam in order to create an array of negative charges. The substrate was then transferred from the SEM chamber to a deposition chamber, where particles were deposited from the gaseous suspension and assembled on charged places of the substrate (Figure 3.16).

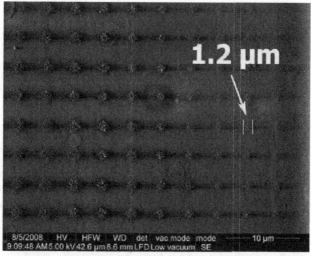

Figure 3.16 Positively charged Pd nanoparticles from gaseous suspension assembled on a negative charge pattern created by an electron beam on a Si_3N_4-coated Si substrate. *Reproduced with permission from Zonnevylle et al. (2009). Copyright by Elsevier.*

Pick-and-place manipulation of microparticles and nanoparticles can be performed using a needle-shaped probe tip and employing various electrostatic effects caused by charging under electron beam illumination. For instance, Ampem-Lassen *et al.* (2009) created a single photon source by such manipulation of a 300-nm diamond nanocrystal in an SEM. The authors used a micromanipulator with a carbon-coated, tapered optical fiber tip attached. When the tip of the micromanipulator approached the diamond nanocrystal sitting on a substrate, the nanocrystal favored the tip (sometimes even jumping from the substrate onto the tip) and could be transported. This pickup effect is believed to be due to the electrostatic interaction resulting from charging in the SEM; however, the exact mechanism was not well explained. Denisyuk *et al.* (2012) and Denisyuk, Komissarenko, & Mukhin (2014) performed some further investigation of this manipulation technique. In particular, Denisyuk, Komissarenko, & Mukhin (2014) reported about pick-and-place manipulation of Al_2O_3, WO_3, and tungsten microparticles and nanoparticles of various shapes (Figure 3.17). This experiment was performed in an SEM employing a nongrounded metalized tip glued to a micromanipulator. The authors demonstrated a well-controlled pickup and drop-off of the particles by the tip. Moreover, the drop-off was demonstrated in two ways: shifting the electron beam from the metallic tip and pulling the tip aside.

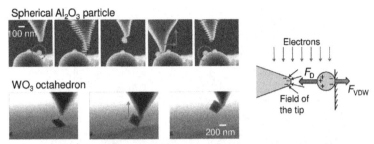

Figure 3.17 Pick-and-place manipulation of an 80-nm Al_2O_3 spherical particle and a WO_3 nanooctahedron in an SEM. The schematic on the right illustrates the mechanism: A particle is retained on a substrate by the van der Waals force, while a nongrounded metallic tip charged under electron beam illumination creates a dielectrophoretic force that pulls the particle from the substrate to the tip. *Reproduced with permission from Denisyuk, Komissarenko, & Mukhin (2014). Copyright by Elsevier.* (See the color plate.)

Denisyuk, Komissarenko, & Mukhin (2014) also created a theoretical model to explain the pickup effect. This model was based on the assumption that particles are retained on the substrate by the van der Waals force, whereas a nongrounded metallic tip gets charged under electron beam illumination and creates electrostatic field gradient and dielectrophoretic forces, which pulls the particle from the substrate to the tip (Figure 3.18). The dielectrophoretic force was computed as an integral over the particle volume:

$$\mathbf{F} = \int (\mathbf{D} - \varepsilon_0 \mathbf{E}) \nabla \mathbf{E} dV, \tag{9}$$

where $\mathbf{E}$ is the local electric field and $\mathbf{D}$ is the local electric displacement field. The tip was supposed to be charged up to the maximum possible value, which was limited by the field emission effect.

Electrostatic manipulation due to the charging effects can be achieved in a TEM as well. However, two cases, discussed next, cannot be considered to be controlled manipulation, but rather, an observed phenomenon of nanoparticle motion induced by some electrostatic mechanism. For example, White *et al.* (2012) studied the dynamics of nanoparticles in a liquid environment. In this experiment, 4–nm Pt nanoparticles were deposited on an insulating membrane of an electron-transparent cell filled with deionized water (Figure 3.19). During imaging the area of 350×350 nm^2 at 300 kV, it appeared that initially immobile particles started to move away from the exposed area. Particle trajectories were directed radially outward from the center, and the dispersion rate increased as the electron beam current

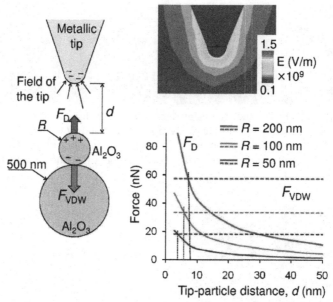

Figure 3.18 Theoretical model of a particle pickup effect: A schematic illustration, distribution of the electric field around the charged tip, and a chart that shows the calculated van der Waals holding force and dielectrophoretic pulling force. *Reproduced with permission from Denisyuk, Komissarenko, & Mukhin (2014). Copyright by Elsevier. (See the color plate.)*

increased from 5 to 57 pA. The observed effect of radially directed motion of the nanoparticles was attributed to the electrophoresis caused by the charging of the membrane and the nanoparticles. White *et al.* (2012) repeated this experiment with a dry sample in a high-vacuum environment under the same imaging conditions, but no motion was detected. Thus, the presence of water reduced the adhesion between the particles and the membrane and allowed the movement of the particles.

The other example of electrostatic motion was observed by Xu *et al.* (2010) in a TEM with normal dry conditions. The authors deposited CdSe nanocrystals on a carbon film from a toluene solution. While imaging of a dried specimen at 200 kV took place, a remarkable effect was detected: one of the thousands of CdSe nanocrystals did not sit on the film; rather, it levitated and slowly rotated (Figure 3.20). The authors suggest that this single 10-nm nanocrystal was trapped in a 3-D Coulomb potential well that was formed due to the charge of the nanocrystal and a unique distribution of two charged rings on the carbon film in a thin-insulating layer that was accidently deposited with the nanocrystals from the solution

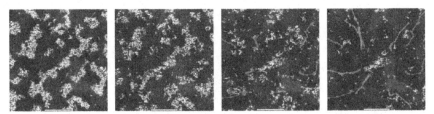

Figure 3.19 Motion of Pt nanoparticles deposited on an insulating membrane of a liquid cell after imaging in a scanning TEM operating at 300 kV with a beam current of 57 pA. The images were taken at 7-s intervals. *Reproduced with permission from White* et al. *(2012). Copyright by American Chemical Society.* (See the color plate.)

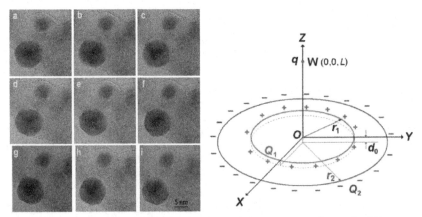

Figure 3.20 A remarkable effect of levitation and slow rotation of a single CdSe nanocrystal over a carbon substrate while imaging in a TEM at 200 kV is shown on the left (the images were captured consequently during 10 minutes of observation). A schematic illustration of the charge distribution, which can cause a 3-D Coulomb potential well responsible for the levitation, is shown at the right. *Reproduced with permission from Xu* et al. *(2010). Copyright by American Chemical Society.* (See the color plate.)

(Figure 3.20). In this case, the Coulomb potential energy of the nanocrystal is given by

$$U(z) = \frac{q}{4\pi\varepsilon_0}\left(\frac{Q_1}{\sqrt{r_1^2 + z^2}} + \frac{Q_2}{\sqrt{r_2^2 + (z + d_0)^2}}\right), \qquad (10)$$

and a potential well responsible for the levitation is produced. Xu *et al.* also suggest that the observed rotational motion of the particle was caused by inelastic interaction with the passing electrons and the fact that the shape and crystalline structure of the particle were somewhat asymmetric. The electromagnetic force caused by passing electrons and its influence on the particle rotation were not considered in this chapter.

5. ELECTROMAGNETIC MANIPULATION

The electromagnetic manipulation technique is based on the influence of the electromagnetic force created by fast electrons. The underlying physics was first considered in the theoretical work by García de Abajo (2004), as described in the section "Electromagnetic Force Exerted on a Particle by Fast-Passing Electrons," earlier in this chapter. The first experimental confirmation of this effect was obtained by Oleshko and Howe (2011) using a TEM.

Due to the similar nature of the optical trapping of particles in optical (or laser) tweezers, such electromagnetic manipulation is known as *electron trapping* or *electron tweezers*. However, in contrast to optical tweezers, which can manipulate submicrometer-sized or larger particles, electron tweezers can operate much smaller particles (with sizes down to 1 nm). Recent developments and progressive use of aberration-corrected TEMs, providing resolutions down to the subangstrom range, further pushed the experimental investigation of this effect. Oleshko and Howe (2013) reviewed electron tweezers in a great deal of detail. Thus, in this review we will just briefly discuss the main experimental results related to this topic.

Oleshko and Howe (2011) described the manipulation of a crystalline Al nanosphere floating in an Al–Si molten alloy bead observed in a TEM at 197 kV (as shown in Figure 3.21). The motion occurred with translation of the beam or with moving the microscope stage. Rotation of the particle under electron illumination was also detected. The effects were attributed to the momentum transfer from the fast electrons to the floating Al nanosphere. The authors reported a rather high value for momentum transfer, which was estimated from the detected motion speed of the nanosphere and was found to be in the range of 10^{-27}–10^{-26} N s. Also, they considered forces exerted on the solid nanoparticle in a liquid environment under electron illumination. The total force is given by

$$\mathbf{F} = \mathbf{F}_d + \mathbf{F}_{gv} + \mathbf{F}_b + \mathbf{F}_r + \mathbf{F}_{gd}, \tag{11}$$

where $\mathbf{F}_d$ is the drag or fluid resistance force, $\mathbf{F}_{gv}$ is the gravitational force, $\mathbf{F}_b$ is the buoyant force, $\mathbf{F}_r$ is the radiation force (due to emitted radiation as a result of the interaction with the electrons), and $\mathbf{F}_{gd}$ is the electromagnetic gradient force induced by the fast electrons. The theoretical value for the electromagnetic force was not provided by Oleshko and Howe (2011).

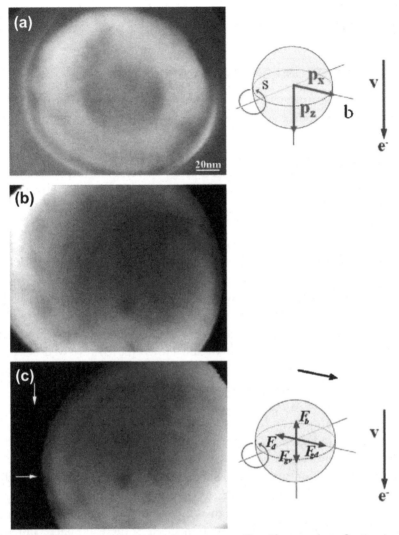

Figure 3.21 TEM image of a 70-nm-diameter crystalline Al nanosphere floating in an Al-Si molten alloy bead (a). Motion of the nanosphere was observed upon translation of the electron beam (b) or upon moving the microscope stage (c). *Reproduced with permission from Oleshko and Howe (2011). Copyright by Elsevier. (See the color plate.)*

Individual nanoparticle manipulation using an electron beam also can be performed in a liquid cell, as reported by Zheng *et al.* (2012). The liquid cell contained a solution with 10-nm gold particles sandwiched between two silicon nitride membranes. For the purpose of manipulation, a 120-kV electron beam was focused on the spot with a Gaussian profile and a diameter

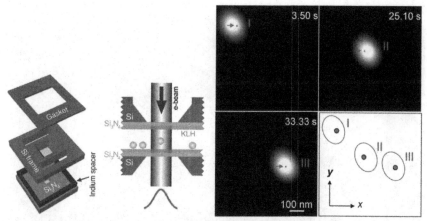

Figure 3.22 Schematic illustration of nanoparticle manipulation inside a liquid cell in a TEM (left), and sequential images demonstrating the motion of an individual gold nanoparticle that follows shifting of the electron beam (right). *Reproduced with permission from Zheng* et al. *(2012). Copyright by American Chemical Society.* (See the color plate.)

in the range of 50–200 nm. An individual particle was then trapped inside the beam and moved following the beam shift (see Figure 3.22). Some Brownian motion of the nanoparticle within the beam was also observed, but the particle did not escape the focusing spot. The authors also estimated that the trapping force produced by the electron beam should be in the order of piconewtons. At the same time, the mechanism of the trapping was not clearly explained, but the authors did discuss that factors such as negative pressure, charging, and thermophoresis caused by the passing electrons might contribute to the effect, but the electromagnetic force induced by the electron beam was not taken into account.

Electromagnetic manipulation of nanoparticles on a substrate in high-vacuum conditions is also possible; however, it was demonstrated only for particles that are a few nanometers in size. For instance, Batson *et al.* (2011 and 2012) reported experimental results on the controlled motion of gold clusters over a carbon substrate in response to 120-kV electrons. The clusters were imaged in a scanning TEM mode with aberration correction. The electron beam stopped at the beginning of each line at the left edge of the scanned area, producing most of the impact on the cluster. The distance between the left edge of the frame and the cluster, called the *impact parameter,* was crucial to the manipulation process. For instance, in the case of a single 1.5-nm cluster and moderate impact parameters, the electron beam caused an attractive force, and so the cluster moved toward

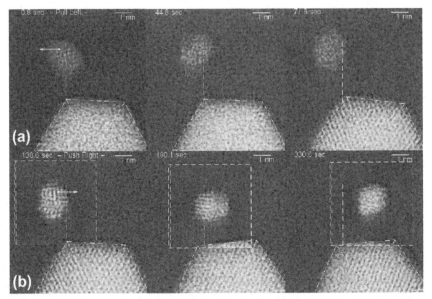

Figure 3.23 Guided motion of a 1.5-nm gold cluster: attractive pulling induced by a moderate impact parameter of 4.5 nm (top), and repulsive pushing resulting from a small impact parameter of 1 nm. *Reproduced with permission from Batson et al. (2011). Copyright by American Chemical Society.* (See the color plate.)

the left edge. On the other hand, for small impact parameters, the force was repulsive and the cluster motion was in the opposite direction (Figure 3.23).

Additional experiments were performed with cluster pairs, where attraction and repulsion between clusters were observed depending on the electron beam placement. The effect of motion was attributed to the influence of an electromagnetic field of the passing electrons, which causes polarization of the particles and induces long-ranged attractive and short-ranged repulsive forces between the electron beam and the particle. The observed effects were explained with the theoretical approach by Reyes-Coronado *et al.* (2010), as discussed in the section "Electromagnetic Force Exerted on a Particle by Fast-Passing Electrons," earlier in this chapter. The authors also note that some additional effects (e.g., specimen charging and heating) may contribute to the motion of the clusters.

Another experiment about the manipulation of nanoparticles on a substrate was performed by Verbeeck *et al.* (2013). Approximately 3-nm gold particles were sitting on a Si_3N_4 supporting film and imaged in a TEM operating at 300 kV. The TEM was equipped with a holographic mask that

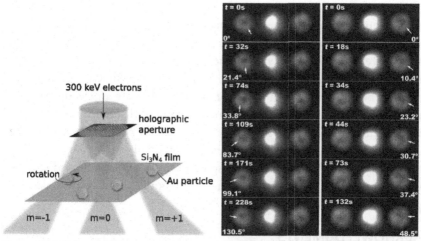

Figure 3.24 Schematic illustration and experimental imaging of particle rotation upon illumination by a focused vortex beam in a TEM. The particle was placed under the beam of $m = -1$ or $m = +1$ orders (marked by the arrows), which caused rotation in different directions. *Reproduced with permission from Verbeeck et al. (2013). Copyright by John Wiley and Sons.* (See the color plate.)

produced a vortex electron beam of $m = -1, 0, +1$ orders, with diameters down to the atomic level. The authors detected the rotation of an individual particle illuminated with a vortex beam. Moreover, the direction of the rotation depended on the diffraction order of the beam (i.e., $m = -1$ or $m = +1$, as shown in Figure 3.24). The observed effect was attributed to the transfer of an angular momentum from the vortex beam to the particle due to the electromagnetic interaction.

6. CONCLUSION

The electron microscope was initially invented as an instrument that could obtain images with a resolution higher than that in an optical microscope. However, a modern electron microscope is not just an imaging tool, but a complex analytical instrument or an advanced nanofabrication workstation. For either of these applications, the possibility for micromanipulation is greatly needed. In this chapter, we described three main methods of using micromanipulation in electron microscopes.

Mechanical (or contact) manipulation involves using an integrated micromanipulator for handling objects while the electron microscope performs imaging. This method was introduced in 1990, and presently it allows not only manipulation with nanoscale precision, but also investigation of several mechanical properties of microobjects and nanoobjects that still are not well known. Other possibilities are electrostatic and electromagnetic noncontact manipulations, which are based on the intrinsic properties of an electron beam to deliver an electrical charge and create an electromagnetic field. The advantage of noncontact manipulation is that it can cause less potential damage to the transported object. Moreover, electromagnetic manipulation, which was experimentally demonstrated only a few years ago, offers the unique potential of precise transportation of nanometer-sized clusters, which is a very interesting phenomenon from both scientific and application points of view.

ACKNOWLEDGMENTS

This work was supported by the government of the Russian Federation (Grant # 074-U01) and the Russian Foundation for Basic Research (Grant # 14-02-31703).

REFERENCES

Ampem-Lassen, E., Simpson, D. A., Gibson, B. C., Trpkovski, S., Hossain, F. M., Huntington, S. T., Ganesan, K., Hollenberg, L. C. L., & Prawer, S. (2009). Nanomanipulation of diamond-based single photon sources. *Optics Express, 17*, 11287–11293.

Ashkin, A. (1970). Acceleration and trapping of particles by radiation pressure. *Physics Review Letters, 24*, 156.

Batson, P. E., Reyes-Coronado, A., Barrera, R. G., Rivacoba, A., Echenique, P. M., & Aizpurua, J. (2011). Plasmonic nanobilliards: Controlling nanoparticle movement using forces induced by swift electrons. *Nano Letters, 11*, 3388–3393.

Batson, P. E., Reyes-Coronado, A., Barrera, R. G., Rivacoba, A., Echenique, P. M., & Aizpurua, J. (2012). Nanoparticle movement: Plasmonic forces and physical constraints. *Ultramicroscopy, 123*, 50–58.

Bergstrom, L. (1997). Hamaker constants of inorganic materials. *Advances in Colloid and Interface Science, 70*, 126–169.

Bussolotti, F., D'Ortenzi, L., Grossi, V., Lozzi, L., Santucci, S., & Passacantando, M. (2007). In situ manipulation and electrical characterization of multiwalled carbon nanotubes by using nanomanipulators under scanning electron microscopy. *Physical Review B, 76*, 125415.

Cagliani, A., Wierzbicki, R., Occhipinti, L., Petersen, D. H., Dyvelkov, K. N., Sukas, Ö. S., Herstrøm, B. G., Booth, T., & Bøggild, P. (2010). Manipulation and in situ transmission electron microscope characterization of sub-100-nm nanostructures using a microfabricated nanogripper. *Journal of Micromechanics and Microengineering, 20*, 035009.

García de Abajo, F. J. (2004). Momentum transfer to small particles by passing electron beams. *Physical Review B., 70*, 115422.

Carpick, R. W., & Salmeron, M. (1997). Scratching the surface: Fundamental investigations of tribology with atomic force microscopy. *Chemical Reviews, 97*, 1163–1194.

Cazaux, J. (2004). Charging in scanning electron microscopy "from inside and outside." *Scanning, 26*, 181–203.

Chen, X., & Wen, J. (2012). In situ wet-cell TEM observation of gold nanoparticle motion in an aqueous solution. *Nanoscale Research Letters, 7*, 598.

Chen, Q., Smith, J. M., Park, J., Kim, K., Ho, D., Rasool, H. I., Zettl, A., & Alivisatos, A. P. (2013). 3D motion of DNA-AU nanoconjugates in graphene liquid cell electron microscopy. *Nano Letters, 13*, 4556–4561.

Clévy, C., Hubert, A., Agnus, J., & Chaillet, N. (2005). A micromanipulation cell including a tool changer. *Journal of Micromechanics and Microengineering, 15*, S292–S301.

Contreras-Naranjo, J. C., & Ugaz, V. M. (2013). A nanometre-scale resolution interference-based probe of interfacial phenomena between microscopic objects and surfaces. *Nature Communications, 4*, 1919.

Cretu, O., Rodríguez-Manzo, J. A., Demortière, A., & Banhart, F. (2012). Electron beam–induced formation and displacement of metal clusters on graphene, carbon nanotubes, and amorphous carbon. *Carbon, 50*, 259–264.

Denisyuk, A. I., Tinskaya, M. A., Petrov, M. I., Shelaev, A. V., & Dorozhkin, P. S. (2012). Tunable optical antennas based on metallic nanoshells with nanoknobs. *Journal of Nanoscience and Nanotechnology, 12*, 8246–8250.

Denisyuk, A. I., Komissarenko, F. E., & Mukhin, I. S. (2014). Electrostatic pick-and-place micro/nanomanipulation under the electron beam. *Microelectric Engineering, 121*, 15–18.

Derjaguin, B. V., Muller, V. M., & Toporov, Y. P. (1975). Effect of contact deformations on the adhesion of particles. *Journal of Colloid Interface Science, 53*, 314–326.

Dong, L., Shou, K., Frutiger, D. R., Subramanian, A., Zhang, L., Nelson, B. J., Tao, X., & Zhang, X. (2008). Engineering multiwalled carbon nanotubes inside a transmission electron microscope using nanorobotic manipulation. *IEEE Transactions on Nanotechnology, 7*, 508–518.

Ducker, W. A., Senden, T. J., & Pashley, R. M. (1991). Direct measurement of colloidal forces using an atomic force microscope. *Nature, 353*, 239–241.

Egerton, R. F., Li, P., & Malac, M. (2004). Radiation damage in the TEM and SEM. *Micron, 35*, 399–409.

Eichhorn, V., Fatikow, S., Wortmann, T., Stolle, C., Edeler, C., Jasper, D., Sardan, O., Bøggild, P., Boetsch, G., Canales, C., & Clavel, R. (2009). NanoLab: A nanorobotic system for automated pick-and-place handling and characterization of CNTs. *Proceedings of IEEE International Conference on Robotics and Automation*, 1826–1831.

Enomoto, K., Kitakata, S., Yasuhara, T., Ohtake, N., Kuzumaki, T., & Mitsuda, Y. (2006). Measurement of Young's modulus of carbon nanotubes by nanoprobe manipulation in a transmission electron microscope. *Applied Physics Letters, 88*, 153115.

Fukata, N., Oshima, T., Tsurui, T., Ito, S., & Murakami, K. (2005). Synthesis of silicon nanowires using laser ablation method and their manipulation by electron beam. *Science and Technology of Advanced Materials, 6*, 628–632.

Fukuda, T., Arai, F., & Dong, L. (2003). Assembly of nanodevices with carbon nanotubes through nanorobotic manipulations. *Proceedings of IEEE, 91*, 1803–1818.

Fukuda, T., Arai, F., & Nakajima, M. (2013). *Micro-nanorobotic Manipulation Systems and Their Applications*. Berlin: Springer.

Götzinger, M., & Peukert, W. (2003). Dispersive forces of particle-surface interactions: Direct AFM measurements and modeling. *Powder Technology, 130*, 102–109.

Grobelny, J., Tsai, J.-H., Kim, D.-I., Pradeep, N., Cook, R. F., & Zachariah, M. R. (2006). Mechanism of nanoparticle manipulation by scanning tunnelling microscopy. *Nanotechnology, 17*, 5519–5524.

Hamaker, H. C. (1937). The London–van der Waals attraction between spherical particles. *Physica, 4*, 1058–1072.

Hänel, K., Birkner, A., Müllegger, S., Winkler, A., & Wöll, C. (2006). Manipulation of organic "needles" using an STM operated under SEM control. *Surface Science, 600*, 2411–2416.

Hatamura, Y., & Morishita, H. (1990). Direct coupling system between nanometer world and human world. *Proceedings of IEEE Micro Electro Mechanical Systems*, 203–208.

Höppener, C., & Novotny, L. (2008). Imaging of membrane proteins using antenna-based optical microscopy. *Nanotechnology, 19*, 384012.

Hu, S., Kim, T. H., Park, J. G., & Busnaina, A. A. (2010). Effect of different deposition mediums on the adhesion and removal of particles. *Journal of the Electrochemical Society, 157*, H662–H665.

Jasper, D. (2011). *SEM-based motion control for automated robotic nanohandling, Dr. rer. nat. dissertation*. Oldenburg, Germany: Carl von Ossietzky University of Oldenburg.

Jasper, D., & Fatikow, S. (2010). Automated high-speed nanopositioning inside scanning electron microscopes. *Proceedings of IEEE Conference on Automation Science and Engineering*, 704–709.

Johnson, K. L., Kendall, K., & Roberts, A. D. (1971). Surface energy and the contact of elastic solids. *Proceedings of the Royal Society of London A, 324*, 301–313.

Kasaya, T., Miyazaki, H. T., Saito, S., Koyano, K., Yamaura, T., & Sato, T. (2004). Image-based autonomous micromanipulation system for arrangement of spheres in a scanning electron microscope. *Review of Scientific Instruments, 75*, 2033–2042.

Kim, K. S., Lim, S. C., Lee, I. B., An, K. H., Bae, D. J., Choi, S., Yoo, J.-E., & Lee, Y. H. (2003). In situ manipulation and characterizations using nanomanipulators inside a field emission-scanning electron microscope. *Review of Scientific Instruments, 74*, 4021–4025.

Kim, S., Shafiei, F., Ratchford, D., & Li, X. (2011). Controlled AFM manipulation of small nanoparticles and assembly of hybrid nanostructures. *Nanotechnology, 22*, 115301.

Krakow, W., & Nixon, W. C. (1977). The behavior of charged particles in the scanning electron microscope. *IEEE Transactions on Industry Applications, IA-13*, 355–366.

Liu, K. H., Gao, P., Xu, Z., Bai, X. D., & Wang, E. G. (2008). In situ probing electrical response on bending of ZnO nanowires inside transmission electron microscope. *Applied Physics Letters, 92*, 213105.

Mai, W., Zhang, L., Gu, Y., Huang, S., Zhang, Z., Lao, C., Yang, P., Qiang, P., & Chen, Z. (2012). Mechanical and electrical characterization of semiconducting ZnO nanorings by direct nano-manipulation. *Applied Physics Letters, 101*, 081910.

Mayer, J., Giannuzzi, L. A., Kamino, T., & Michael, J. (2007). TEM sample preparation and FIB-induced damage. *MRS Bulletin, 32*, 400–407.

Min, Y., Akbulut, M., Kristiansen, K., Golan, Y., & Israelachvili, J. (2008). The role of interparticle and external forces in nanoparticle assembly. *Nature Materials, 7*, 527–538.

Miyazaki, H., & Sato, T. (1996). Pick-and-place shape forming of three-dimensional micro structures from fine particles. *Proceedings of IEEE*, 2535–2540.

Miyazaki, H. T., Tomizawa, Y., Koyano, K., Sato, T., & Shinya, N. (2000a). Adhesion force measurement system for micro-objects in a scanning electron microscope. *Review of Scientific Instruments, 71*, 3123–3131.

Miyazaki, H. T., Tomizawa, Y., Saito, S., Sato, T., & Shinya, N. (2000b). Adhesion of micrometer-sized polymer particles under a scanning electron microscope. *Journal of Applied Physics, 88*, 3330–3340.

Mølhave, K., Wich, T., Kortschack, A., & Bøggild, P. (2006). Pick-and-place nanomanipulation using microfabricated grippers. *Nanotechnology, 17*, 2434–2441.

Nakajima, M., Arai, F., & Fukuda, T. (2006). In situ measurement of Young's modulus of carbon nanotubes inside a TEM through a hybrid nanorobotic manipulation system. *IEEE Transactions on Nanotechnology, 5*, 243–248.

Nakazato, Y., Yuasa, T., Sekine, G., Miyazawa, H., Jin, M., Takeuchi, S., Ariga, Y., & Murakawa, M. (2009). Micromanipulation system using scanning electron microscope. *Microsystems Technology, 15*, 859–864.

Neuman, K. C., & Block, S. M. (2004). Optical trapping. *Review of Scientific instruments, 75*, 2787–2809.

Niguès, A., Siria, A., Vincent, P., Poncharal, P., & Bocquet, L. (2014). Ultrahigh interlayer friction in multiwalled boron nitride nanotubes. *Nature Materials, 13,* 688–693.

Oleshko, V. P., & Howe, J. M. (2011). Are electron tweezers possible? *Ultramicroscopy, 111,* 1599–1606.

Oleshko, V. P., & Howe, J. M. (2013). Electron tweezers as a tool for high-precision manipulation of nanoobjects. *Advances in Imaging and Electron Physics, 179,* 203–262.

Peng, L.-M., Chen, Q., Liang, X. L., Gao, S., Wang, J. Y., Kleindiek, S., & Taic, S. W. (2004). Performing probe experiments in the SEM. *Micron, 35,* 495–502.

Polyakov, B., Dorogin, L. M., Vlassov, S., Kink, I., Lohmus, A., Romanov, A. E., & Lohmus, R. (2011). Real-time measurements of sliding friction and elastic properties of ZnO nanowires inside a scanning electron microscope. *Solid State Communications, 151,* 1244–1247.

Polyakov, B., Vlassov, S., Dorogin, L. M., Kulis, P., Kink, I., & Lohmus, R. (2012). The effect of substrate roughness on the static friction of CuO nanowires. *Surface Science, 606,* 1393–1399.

Polyakov, B., Vlassov, S., Dorogin, L. M., Novoselska, N., Butikova, J., Antsov, M., Oras, S., Lohmus, R., & Kink, I. (2014). Some aspects of formation and tribological properties of silver nanodumbbells. *Nanoscale Research Letters, 9,* 186.

Reyes-Coronado, A., Barrera, R. G., Batson, P. E., Echenique, P. M., Rivacoba, A., & Aizpurua, J. (2010). Electromagnetic forces on plasmonic nanoparticles induced by fast electron beams. *Physical Review B, 82,* 235429.

Ru, C., & To, S. (2012b). Contact detection for nanomanipulation in a scanning electron microscope. *Ultramicroscopy, 118,* 61–66.

Ru, C., Zhang, Y., Huang, H., & Chen, T. (2012a). An improved visual tracking method in scanning electron microscope. *Microscopy and Microanalysis, 18,* 612–620.

Saito, S., Miyazaki, H. T., Sato, T., & Takahashi, K. (2002). Kinematics of mechanical and adhesional micromanipulation under a scanning electron microscope. *Journal of Applied Physics, 92,* 5140–5149.

Sitti, M., & Hashimoto, H. (2000). Controlled pushing of nanoparticles: Modeling and experiments. *IEEE/ASME Transactions of Mechatronics, 5,* 199–211.

Van Huis, M. A., Kunneman, L. T., Overgaag, K., Xu, Q., Pandraud, G., Zandbergen, H. W., & Vanmaekelbergh, D. (2008). Low-temperature nanocrystal unification through rotations and relaxations probed by in situ transmission electron microscopy. *Nano Letters, 8,* 3959–3963.

Verbeeck, J., Tian, H., & Van Tendeloo, G. (2013). How to manipulate nanoparticles with an electron beam? *Advanced Materials, 25,* 1114–1117.

Vlassov, S., Polyakov, B., Dorogin, L. M., Lõhmus, A., Romanov, A. E., Kink, I., Gnecco, E., & Lõhmus, R. (2011). Real-time manipulation of gold nanoparticles inside a scanning electron microscope. *Solid State Communications, 151,* 688–692.

Vlassov, S., Polyakov, B., Dorogin, L. M., Antsov, M., Mets, M., Umalas, M., Saar, R., Lõhmus, R., & Kink, I. (2014). Elasticity and yield strength of pentagonal silver nanowires: In situ bending tests. *Materials Chemistry and Physics, 143,* 1026–1031.

White, E. R., Mecklenburg, M., Shevitski, B., Singer, S. B., & Regan, B. C. (2012). Charged nanoparticle dynamics in water induced by scanning transmission electron microscopy. *Langmuir, 28,* 3695–3698.

Williams, P. (1987). Motion of small gold clusters in the electron microscope. *Applied Physics Letters, 50,* 1760–1762.

Xu, S.-Y., Sun, W.-Q., Zhang, M., Xu, J., & Peng, L.-M. (2010). Transmission electron microscope observation of a freestanding nanocrystal in a Coulomb potential well. *Nanoscale, 2,* 248–253.

Ye, X. (2012). *Towards automated nanomanipulation under scanning electron microscopy, Master of science thesis.* Toronto, Canada: University of Toronto.

Ye, X., Zhang, Y., Ru, C., Luo, J., Xie, S., & Sun, Y. (2013). Automated pick-place of silicon nanowires. *IEEE Transactions on Automation Science and Engineering*, 1–8.

Yu, M., Dyer, M. J., Skidmore, G. D., Rohrs, H. W., Lu, X., Ausman, K. D., Von Ehr, J. R., & Ruoff, R. S. (1999). Three-dimensional manipulation of carbon nanotubes under a scanning electron microscope. *Nanotechnology, 10*, 244–252.

Zhang, Y. L., Li, J., To, S., Zhang, Y., Ye, X., You, L., & Sun, Y. (2012). Automated nanomanipulation for nanodevice construction. *Nanotechnology, 23*, 065304.

Zhang, Y. L., Zhang, Y., Ru, C., Chen, B. K., & Sun, Y. (2013). A load-lock-compatible nanomanipulation system for scanning electron microscope. *IEEE/ASME Transactions on Mechatronics, 18*, 230–237.

Zheng, Q., Jiang, B., Liu, S., Weng, Y., Lu, L., Xue, Q., Zhu, J., Jiang, Q., Wang, S., & Peng, L. (2008). Self-retracting motion of graphite microflakes. *Physics Review Letters, 100*, 067205.

Zheng, H., Mirsaidov, U. M., Wang, L.-W., & Matsudaira, P. (2012). Electron beam manipulation of nanoparticles. *Nano Letters, 12*, 5644–5648.

Zonnevylle, A. C., Hagen, C. W., Kruit, P., & Schmidt-Ott, A. (2009). Directed assembly of nano-particles with the help of charge patterns created with scanning electron microscope. *Microelectric Engineering, 86*, 803–805.

CONTENTS OF VOLUMES 151–185

---

[1] Lists of the contents of volumes 100–149 are to be found in volume 150; the entire series can be searched on ScienceDirect.com

INDEX

Note: Page numbers followed by "f" and "t" indicate figures and tables respectively.

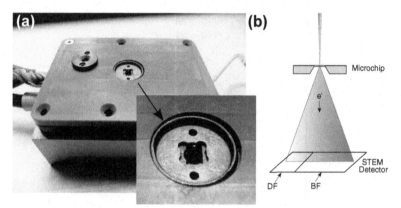

PLATE 1 (Figure 1.8. on page 14 of this Volume)

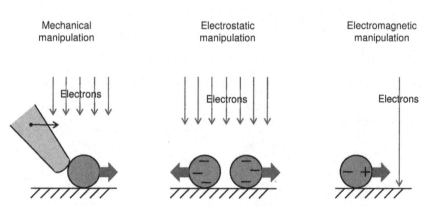

PLATE 2 (Figure 3.1. on page 103 of this Volume)

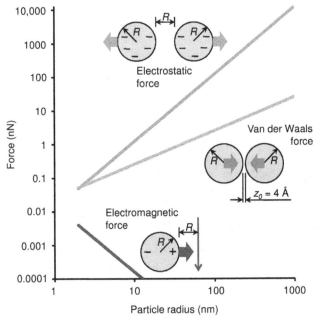

PLATE 3 (Figure 3.2. on page 105 of this Volume)

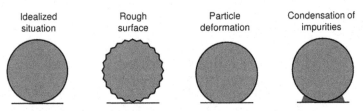

PLATE 4 (Figure 3.4. on page 107 of this Volume)

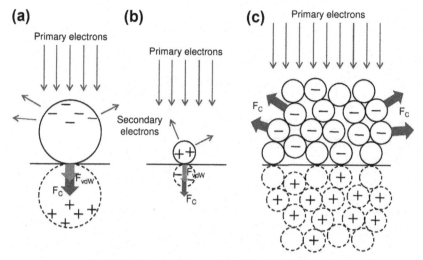

PLATE 5 (Figure 3.5. on page 108 of this Volume)

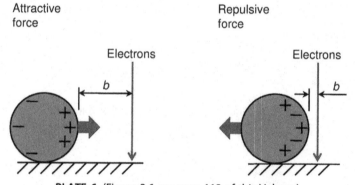

PLATE 6 (Figure 3.6. on page 112 of this Volume)

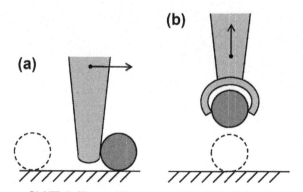

PLATE 7 (Figure 3.7. on page 114 of this Volume)

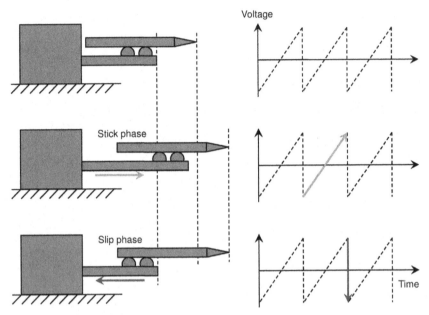

PLATE 8 (Figure 3.8. on page 116 of this Volume)

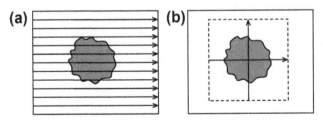

PLATE 9 (Figure 3.9. on page 119 of this Volume)

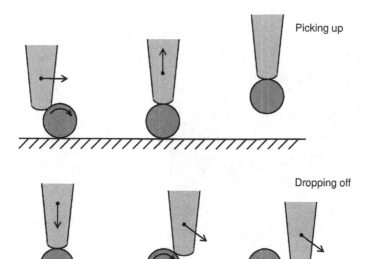

Picking up

Dropping off

PLATE 10 (Figure 3.11. on page 122 of this Volume)

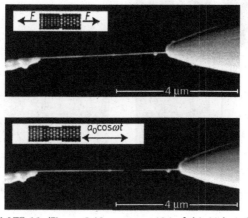

PLATE 11 (Figure 3.13. on page 124 of this Volume)

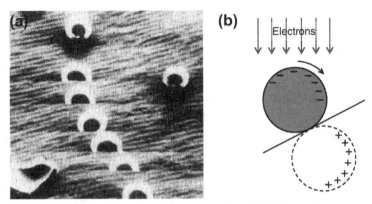

PLATE 12 (Figure 3.14. on page 125 of this Volume)

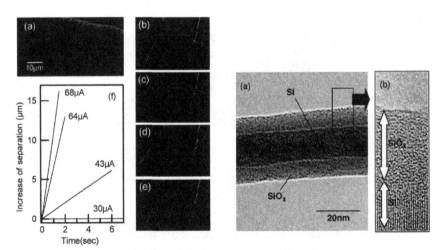

PLATE 13 (Figure 3.15. on page 126 of this Volume)

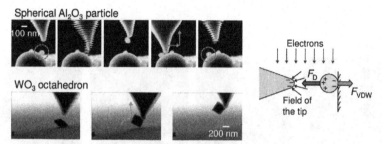

PLATE 14 (Figure 3.17. on page 128 of this Volume)

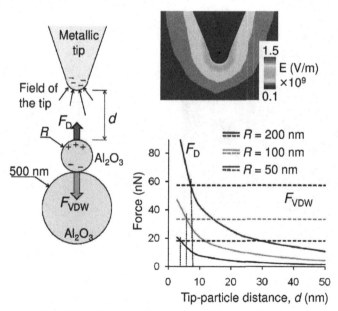

PLATE 15 (Figure 3.18. on page 129 of this Volume)

PLATE 16 (Figure 3.19. on page 130 of this Volume)

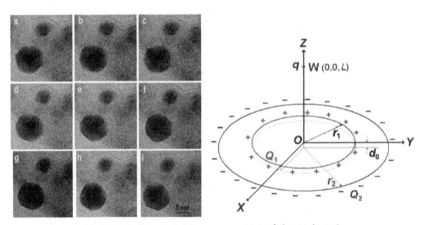

PLATE 17 (Figure 3.20. on page 130 of this Volume)

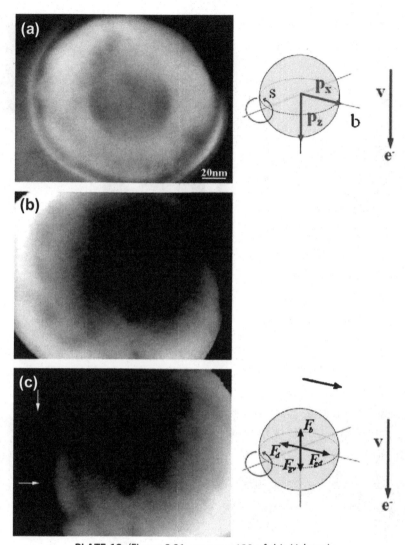

PLATE 18 (Figure 3.21. on page 132 of this Volume)

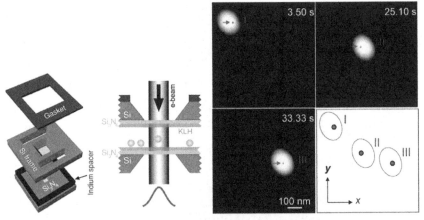

PLATE 19 (Figure 3.22. on page 133 of this Volume)

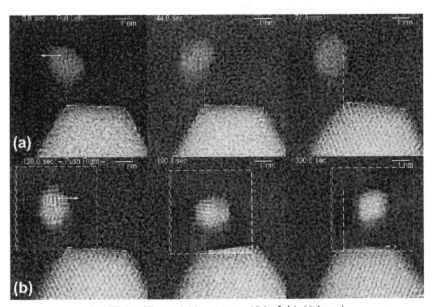

PLATE 20 (Figure 3.23. on page 134 of this Volume)

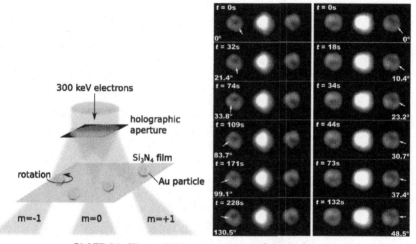

PLATE 21 (Figure 3.24. on page 135 of this Volume)

Printed in the United States
By Bookmasters